General System Theory

Foundations, Development, Applications

General System Theory

Foundations, Development, Applications

Revised Edition

by Ludwig von Bertalanffy

GEORGE BRAZILLER
New York

For information address the publisher,
George Braziller, Inc.
One Park Avenue
New York, N.Y. 10016

Fourth Printing, April 1973
Standard Book Number: 0–8076–0452–6, cloth
                                    0–8076–0453–4, paper
Library of Congress Catalog Card Number: 68–25176
Printed in the United States of America

MANIBUS *Nicolai de Cusa Cardinalis, Gottfriedi Guglielmi Leibnitii, Joannis Wolfgangi de Goethe Aldique Huxleyi, necnon de Bertalanffy Pauli, S.J., antecessoris, cosmographi*

Foreword

The present volume appears to demand some introductory notes clarifying its scope, content, and method of presentation.

There is a large number of texts, monographs, symposia, etc., devoted to "systems" and "systems theory". "Systems Science," or one of its many synonyms, is rapidly becoming part of the established university curriculum. This is predominantly a development in engineering science in the broad sense, necessitated by the complexity of "systems" in modern technology, man-machine relations, programming and similar considerations which were not felt in yesteryear's technology but which have become imperative in the complex technological and social structures of the modern world. Systems theory, in this sense, is preeminently a mathematical field, offering partly novel and highly sophisticated techniques, closely linked with computer science, and essentially determined by the requirement to cope with a new sort of problem that has been appearing.

What may be obscured in these developments—important as they are—is the fact that systems theory is a broad view which far transcends technological problems and demands, a reorientation that has become necessary in science in general and in the gamut of disciplines from physics and biology to the behavioral and social sciences and to philosophy. It is operative, with varying degrees of success and exactitude, in various realms, and heralds a new world view of considerable impact. The student in "systems science" receives a technical training which makes systems theory—originally intended to overcome current overspecialization—into another of the hundreds of academic specialties. Moreover, systems science, centered in computer technology, cybernetics, automation and systems engineering, appears to make the systems idea another—and indeed the ultimate—technique to shape man and society ever more into the "megamachine" which Mumford (1967) has so impressively described in its advance through history.

The present book hopes to make a contribution in both respects implied in the above: offering to the student of systems science a broadened perspective, and to the general reader a panoramic view of a development which is indubitably characteristic of and important in the present world. While fully realizing his limitations and shortcomings, the author feels entitled to do so because he was among the first to introduce general system theory, which is now becoming an important field of research and application.

As Simon (1965) correctly remarked, an introduction into a rapidly developing field largely consists in its conceptual history. It may not be inappropriate, therefore, that the present work consists of studies written over a period of some thirty years. The book thus presents systems theory not as a rigid doctrine (which at present it is not) but rather in its becoming and in the development of its ideas which, hopefully, can serve as a basis for further study and investigation.

In order to serve the purpose, these studies were arranged in logical rather than chronological order and were carefully edited. Editing was limited, however, to elimination of repetitions, minor stylistic improvements and some suitable rearrangements. Intentionally, no changes in content were made from hindsight gained at a later time. Repetitions could not be completely avoided because similar ideas sometimes appeared in different contexts; but it is hoped they were kept at a tolerable level. They may even be not undesirable to the student seeking the general idea or its application in a specific field.

The original sources are indicated in the list of Acknowledgments. For evaluation of the material presented and reasons of priority which will become apparent, some major data may be summarized as follows. Chapter 5 (1940) introduced the "theory of the organism as open system." Together with Burton's (1939) work, this was the original statement of the concept which gained increasing importance and application. This publication remained almost unknown among British and American scientists and is therefore reproduced in its entirety, although much can be added, as is partly reviewed in Chapters 7 (1964) and 6 (1967). Similarly, the first announcement of general system theory

(1945) is reproduced as Chapter 3, abridged and somewhat re-arranged, but otherwise true to the original. The Appendix (review of an address presented in 1947) is reproduced as an early statement long before systems theory and cognate terms and fields appeared academically or in technology. A review in nontechnical language (1956) serves as Chapter 2; Chapters 1 and 4 try to bring the story up to date.

The author wishes to extend his thanks to many persons and agencies that facilitated the work here presented. Thanks are due to Dr. George Brantl, editor at George Braziller, Inc., for having suggested the publication and for his valuable editorial assistance in presenting the book to its advantage. The permissions of editors and publishers where the essays were first published, as indicated in the source list, are gratefully acknowledged. So is the assistance of various agencies, the National Research Council and National Cancer Institute of Canada, the Canada Council, the University of Alberta General Research Committee and others, which sponsored part of the work here reported by research grants and other support. The author's secretary, Mrs. Elizabeth Grundau, took care of the manuscript in its various phases, assisted in bibliographic and library work, and provided translations of the chapters originally published in German, thus far exceeding secretarial routine. Last but not least, my wife, Maria von Bertalanffy, has to be thanked for her untiring help and criticism when these essays were written. Without the encouragement of colleagues, too numerous to mention, the writer, in the face of obstructions and obstacles, would hardly have persevered in the task of introducing and developing general system theory.

L.v.B.

University of Alberta
Edmonton (Canada)
March 1968

Acknowledgments

Most of the chapters in this volume have previously appeared, sometimes in modified form. The publication history is given here for each chapter. The author wishes to thank the original publishers of the articles for permission to include them in this volume:

Chapter 1: Written for this volume (1967).

Chapter 2: "General System Theory," in *Main Currents in Modern Thought*, Volume 11, #4, March 1955, pp. 75–83. Reprinted in *General Systems*, 1 (1956) 1–10; R. W. Taylor (ed.), *Life, Language, Law*: Essays in Honor of A. F. Bentley, Yellow Springs (Ohio), Antioch Press, 1957, pp. 58–78; J. D. Singer (ed.), *Human Behavior and International Politics*, Chicago, Rand McNally & Co., 1965, pp. 20–31; N. J. Demerath III and R. A. Peterson (eds.), *System, Change, and Conflict*, Glencoe (Ill.), Free Press, 1967. Additions were taken from "Allgemeine Systemtheorie. Wege zu einer neuen Mathesis universalis," *Deutsche Universitätszeitung*, 5/6 (1957) 8–12. Also in Italian, "La Teoria generale dei Sistemi," *La Voce dell' America*, 18-G and 2-H (1956–57), and French, "Histoire et méthodes de la théorie générale des systèmes," *Atomes*, 21 (1966) 100–104.

Chapter 3: Condensed from "Zu einer allgemeinen Systemlehre," *Deutsche Zeitschrift für Philosophie*, 18, No. 3/4 (1945); "An Outline of General System Theory," *British Journal of the Philosophy of Science*, 1 (1950) 139–164; "Zu einer allgemeinen Systemlehre," *Biologia Generalis*, 19 (1949) 114–129.

Chapter 4: "General System Theory. A Critical Review." *General Systems*, 7 (1962), 1–20. Reprinted in W. Buckley (ed.), *Modern Systems Research for the Behavioral Scientist*, Chicago, Aldine Publishing Co., 1968, pp. 11–30.

Chapter 5: "Der Organismus als physikalisches System betrachtet," *Die Naturwissenschaften*, 28 (1940) 521–531.

Chapter 6: "Das Modell des offenen Systems," *Nova Acta Leopoldina*, (1969).

Chapter 7: "Basic Concepts in Quantitative Biology of Metabolism," *Helgoländer Wissenschaftliche Meeresuntersuchungen*, 9 (First International Symposium on Quantitative Biology of Metabolism) (1964) 5–37.

Chapter 8: Substance of lectures presented at the University of Western Ontario (London), University of California Medical School (San Francisco), University of Alberta (Edmonton, Calgary), etc., 1961–64.

Chapter 9: "General System Theory and Psychiatry," from Chapter 43 of *The American Handbook of Psychiatry*, Vol. 3, edited by Silvano Arieti, © 1966 by Basic Books, Inc., Publishers, New York.

Chapter 10: "An Essay on the Relativity of Categories," *Philosophy of Science*, 22 (1955) 243–263. Reprinted in *General Systems*, 7 (1962) 71–83.

Appendix "Vom Sinn und der Einheit der Naturwissenschaften. Aus einem Vortrag von Prof. Dr. Ludwig von Bertalanffy," *Der Student* (Wien), 2, No. 7/8 (1947) 10–11.

Contents

Foreword vii

Acknowledgments xi

Preface to the Revised Edition xvii

1 Introduction 3
    Systems Everywhere 3
    On the History of Systems Theory 10
    Trends in Systems Theory 17

2 The Meaning of General System Theory 30
    The Quest for a General System Theory 30
    Aims of General System Theory 36
    Closed and Open Systems: Limitations of
        Conventional Physics 39
    Information and Entropy 41
    Causality and Teleology 44
    What Is Organization? 46
    General System Theory and the Unity of Science 48
    General System Theory in Education: The
        Production of Scientific Generalists 49
    Science and Society 51
    The Ultimate Precept: Man as the Individual 52

3 Some System Concepts in Elementary Mathematical
  Consideration 54
    The System Concept 54
    Growth 60
    Competition 63
    Wholeness, Sum, Mechanization, Centralization 66
    Finality 75
    Types of Finality 77
    Isomorphism in Science 80
    The Unity of Science 86

4   Advances in General System Theory 87

      Approaches and Aims in Systems Science 87

      Methods in General Systems Research 94

      Advances of General System Theory 99

5   The Organism Considered as Physical System 120

      The Organism as Open System 120

      General Characteristics of Open

        Chemical Systems 124

      Equifinality 131

      Biological Applications 134

6   The Model of Open System 139

      The Living Machine and Its Limitations 139

      Some Characteristics of Open Systems 141

      Open Systems in Biology 145

      Open Systems and Cybernetics 149

      Unsolved Problems 151

      Conclusion 153

7   Some Aspects of System Theory in Biology 155

      Open Systems and Steady States 156

      Feedback and Homeostasis 160

      Allometry and the Surface Rule 163

      Theory of Animal Growth 171

      Summary 184

8   The System Concept in the Sciences of Man 186

      The Organismic Revolution 186

      The Image of Man in Contemporary Thought 188

      System-Theoretical Re-orientation 192

      Systems in the Social Sciences 194

      A System-Theoretical Concept of History 197

      The Future in System-Theoretical Aspect 203

9   General System Theory in Psychology and Psychiatry 205

      The Quandary of Modern Psychology 205

      System Concepts in Psychopathology 208

      Conclusion 220

10   The Relativity of Categories 222

      The Whorfian Hypothesis 222

      The Biological Relativity of Categories 227

The Cultural Relativity of Categories 232

The Perspectivistic View 239

Notes 248

Appendix I: Notes on Developments in
Mathematical System Theory 251

Appendix II: The Meaning and Unity of Science 257

References 260

Suggestions for Further Reading 281

Index 285

Preface to the Revised Edition

In the few years since this book was first published considerable developments in general system theory have taken place. So I welcome the opportunity presented by this revised edition to offer a few comments from the present point of advantage.

The postulate and term of general system theory was introduced by me some thirty years ago. Since then general system theory—under this or similar denominations—has become a recognized discipline, with university courses, texts, books of readings, journals, meetings, working groups, centers, and other *accoutrements* of an academic field of teaching and research. Thus the postulate of a "new science" which I had presented, has become a reality.

This was based on many developments which will be reviewed in the present book. The system viewpoint has penetrated, and has indeed proved indispensable, in a vast variety of scientific and technological fields. This, and the further fact that it represents a novel "paradigm" in scientific thinking (to use Thomas Kuhn's expression), has as a consequence that the concept of system can be defined and developed in different ways as required by the objectives of research, and as reflecting different aspects of the central notion.

Under these circumstances an introduction into the field is possible in two ways. One can either accept one of the available models and definitions of system and rigorously derive the consequent theory. Such presentations are fortunately available; some of them will be cited in the following.

The other approach—which is followed in the present book—is to start from problems as they have arisen in the various sciences, to show the necessity of the system viewpoint, and to develop it, in more or less detail, in a selection of illustrative examples. Such procedure will not present a rigorous development of theory and the examples used will be replaceable, that is, others and possibly better ones might be given in the way of illustration. It is, however, the writer's experience—and, judging from the wide acceptance the present book has found, also the experience of others—that such panoramic view presents to the student a suitable introduction into a new way of thinking which is eagerly and even enthusiastically accepted; and offers to the advanced scholar a starting point for further work. The latter is testified by the large number of investigations which drew their inspiration from the present work.

A competent critic (Robert Rosen in *Science*, 164, 1969, p. 681) found "surprisingly few anachronisms in need of correction" in the present book, even though some of the chapters contained in it go back 30 years. This is high praise considering the fact that nowadays scientific monographs frequently are very much "in need of correction" even at the time of appearance. This was not, as the reviewer implied, the result of clever editing (editing was in fact limited to a bare minimum of stylistic improvement) but apparently the author had been "right," in the sense that he had laid a sound basis and correctly forecast future developments. One may, for example, look at systems problems as listed in the paragraph, "Isomorphism in Science" of the present book; these (and others) are presently being answered by dynamical system theory and control theory. The isomorphism of laws is shown in this book by examples which were chosen as intentionally simple illustrations; but the same applies for more sophisticated cases which are far from being mathematically trivial. For example, . . . it is a striking fact that biological systems as diverse as the central nervous system, and the biochemical regulatory network in cells should be strictly analogous . . . it is all the more remarkable when it is realized that this particular analogy between different systems at different levels of biological organization is but one member of a large class of such analogies. (Rosen, 1967.)

On a more general level, the "parallelism of general cognitive principles in different fields" was noted in the present volume in a number of instances. But it was not then foreseen that general system theory would play quite an important role in modern orientations in geography or that it parallels French structuralism (e.g. Piaget, Lévi-Strauss) and would exert considerable influence on American functionalism in sociology.

With the increasing expansion of systems thinking and studies, the definition of general system theory came under renewed scrutiny. Some indication as to its meaning and scope may therefore be in place. The term "general system theory" was introduced by the present author, deliberately, in a catholic sense. One may, of course, limit it to the "technical" meaning in the sense of mathematical theory (as is frequently done), but this appears unadvisable in view of the fact that there are many "system" problems asking for "theory" which latter is not at present available in mathematical terms. So the name "general system theory" is here used broadly, similar to our speaking of the "theory of evolution," which comprises about everything between fossil digging, anatomy and the mathematical theory of selection; or "behaviour theory" extending from bird watching to sophisticated neurophysiological theories. It is the introduction of a new paradigm that matters.

Broadly speaking, three main aspects can be indicated which are not separable in content but distinguishable in intention. The first may be circumscribed as *"systems science,"* i.e. scientific exploration and theory of "systems" in the various sciences (e.g. physics, biology, psychology, social sciences), and general system theory as doctrine of principles applying to all (or defined subclasses of) systems.

Entities of an essentially new sort are entering the sphere of scientific thought. Classical science in its diverse disciplines, be it chemistry, biology, psychology or the social sciences, tried to isolate the elements of the observed universe—chemical compounds and enzymes, cells, elementary sensations, freely competing individuals, what not—expecting that, by putting them together again, conceptually or experimentally, the whole or system—cell, mind, society—would result and be intelligible. Now we have learned that for an understanding not only the elements but their interrelations as well are required: say, the interplay of enzymes in a cell, of many mental processes conscious and unconscious, the structure and dynamics of social systems and the like. This requires exploration of the many systems in our observed universe in their own right and specificities. Furthermore, it turns out that there are general aspects, correspondences and isomorphisms common to "systems." This is the domain of *general system theory;* indeed, such parallelisms or isomorphies appear—sometimes surprisingly—in otherwise totally different "systems." General system theory, then, is scientific exploration of "wholes" and "wholeness" which, not so long ago, were considered to be metaphysical notions transcending the boundaries of science. Novel conceptions, models and mathematical fields have developed to deal with them, such as dynamical system theory, cybernetics, automata theory, system analysis by set, net, graph theory and others.

The second realm is *"systems technology,"* that is, the problems arising in modern technology and society, comprising both the "hardware" of computers, automation, self-regulating machinery, etc., and the "software" of new theoretical developments and disciplines.

Modern technology and society have become so complex that traditional ways and means are not sufficient any more but approaches of a holistic or systems, and generalist or inter-disciplinary nature became necessary. This is true in many ways. Systems of many levels ask for scientific control: ecosystems the disturbance of which results in pressing problems like pollution; formal organizations like a bureaucracy, educational institution or army; the grave problems appearing in socio-economic systems, in international relations, politics and deterrence. Irrespective of the questions of how far scientific understanding (contrasted to the admission of irrationality of cultural and historical events) is possible, and to what extent scientific control is feasible or even desirable, there can be no dispute that these are essentially "systems" problems, that is, problems of interrelations of a great number of "variables." The same applies to narrower objectives in industry, commerce and armament. The technological demands have led to novel conceptions and disciplines, partly of great originality and introducing new basic notions, such as control and information theory, game, decision theory, theory of circuits and queueing, etc. The general characteristic, again, is that these were the offspring of specific and concrete problems in technology, but models, conceptualization and principles—as, for example, the concepts of information, feedback, control, stability, circuit theory, etc.—by far transcended specialist boundaries, were of an inter-disciplinary nature, and were found to be independent of their special realizations, as exemplified by isomorphic feedback models in mechanical, hydrodynamic, electrical, biological, etc., systems. Similarly, developments originating in pure and in applied science converge, as in dynamical system theory and control theory.

Again, there is a spectrum from highly sophisticated mathematical theory, to computer simulation where variables can be treated quantitatively but analytical solutions are lacking, to more or less informal discussion of problems of a system nature.

Thirdly, there is *systems philosophy*, i.e. the reorientation of thought and world view ensuing from the introduction of "system" as a new scientific paradigm (in contrast to the analytic, mechanistic, one-way causal paradigm of classical science). As every scientific theory of broader scope, general system theory has its "metascientific" or philosophical aspects. The concept of "system" constitutes a new "paradigm," in Thomas Kuhn's phrase, or as the present writer (1967) put it, a "new philosophy of nature," contrasting the "blind laws of nature" of the mechanistic world view and the world process as a Shakespearean tale told by an idiot, with an organismic outlook of the "world as a great organization."

This essentially divides into three parts. First, we must find out the "nature of the beast." This is *systems ontology*—what is meant by "system" and how systems are realized at the various levels of the world of observation.

What is to be defined and described as system is not a question with an obvious or trivial answer. It will be readily agreed that a galaxy, a dog, a cell and an atom are *real systems*; that is, entities perceived in or inferred from observation, and existing independently of an observer. On the other hand, there are *conceptual systems* such as logic, mathematics (but e.g. also including music) which essentially are symbolic constructs; with *abstracted systems* (science) as a subclass of the latter, i.e. conceptual systems corresponding with reality.

However, the distinction is by no means as sharp and clear as it would appear. An ecosystem or social system is "real" enough, as we uncomfortably experience when, for example, the ecosystem is disturbed by pollution, or society presents us with so many unsolved problems. But these are not objects of perception or direct observation; they are conceptual constructs. The same is true even of the objects of our everyday world which by no means are simply "given" as sense data or simple perceptions, but actually are construed by an enormity of "mental" factors ranging from gestalt dynamics and learning processes to linguistic and cultural factors largely determining what we actually "see" or perceive. Thus the distinction between "real" objects and systems as given in observation and "conceptual" constructs and systems cannot be drawn in any commonsense way. These are deep problems which can only be indicated in this context.

This leads to *systems epistemology*. As is apparent from the above, this is profoundly different from the epistemology of logical positivism or empiricism even though it shares their scientific attitude. The epistemology (and metaphysics) of logical positivism was determined by the ideas of physicalism, atomism, and the "camera-theory" of knowledge. These, in view of present-day knowledge, are obsolete. As against physicalism and reductionism, the problems and modes of thought occurring in the biological, behavioral and social sciences require equal consideration and simple "reduction" to the elementary particles and conventional laws of physics does not appear feasible. Compared to the analytical procedure of classical science with resolution into component elements and one-way or linear causality as basic category, the investigation of organized wholes of many variables requires new categories of interaction, transaction, organization, teleology, etc., with many problems arising for epistemology, mathematical models and techniques. Furthermore, perception is not a reflection of "real things" (whatever their metaphysical status), and knowledge not a simple approximation to "truth" or "reality." It is an interaction between knower and known, this dependent on a multiplicity of factors of a biological, psychological, cultural, linguistic, etc., nature. Physics itself tells that there are no ultimate entities like corpuscles or waves, existing independent of the observer. This leads to a "perspective" philosophy for which physics, fully acknowledging its achievements in its own and related fields, is not a monopolistic way of knowledge. Against reductionism and theories declaring that reality is "nothing but" (a heap of physical particles, genes, reflexes, drives, or whatever the case may be), we see science as one of the "perspectives" man with his biological, cultural and linguistic endowment and bondage, has created to deal with the universe he is "thrown in," or rather to which he is adapted owing to evolution and history.

The third part of systems philosophy will be concerned with the relations of man and world or what is termed *"values"* in philosophical parlance. If reality is a hierarchy of organized wholes, the image of man will be different from what it is in a world of physical particles governed by chance events as ultimate and only "true" reality. Rather, the world of symbols, values, social entities and cultures is something very "real"; and its embeddedness in a cosmic order of hierarchies is apt to bridge the opposition of C. P. Snow's "Two Cultures" of science and the humanities, technology and history, natural and social sciences, or in whatever way the antithesis is formulated.

This humanistic concern of general system theory as I understand it makes a difference to mechanistically oriented system theorists speaking solely in terms of mathematics, feedback and technology and so giving rise to the fear that system theory is indeed the ultimate step towards mechanization and devaluation of man and towards technocratic society. While understanding and emphasizing the aspect of mathematics, pure and applied science, I do not see that these humanistic aspects can be evaded if general system theory is not limited to a restricted and fractional vision.

Here is perhaps another reason for using this book as an introduction into the field. A textbook-like presentation must follow the straight and narrow path of mathematical and scientific rectitude. The necessity of such "technical" exposition needs no elaboration. But there are many more problems encompassed in general system theory to which the present volume will lead.

The volume contains, besides a rather comprehensive bibliography indicating the sources cited in the text, a list of suggested readings which the student may find useful. More specifically, the following recent publications will be found to be valuable texts on topics introduced in the present book. Discussion of the various approaches in general system theory is found in *Trends in General Systems Theory* (G. Klir, ed.) and "Unity through Diversity" (*Festschrift in Honor of L. von Bertalanffy*, W. Gray and N. Rizzo, eds.), especially Books II and IV. *Dynamical System Theory* is developed in the book of this title by Robert Rosen. An excellent presentation of dynamical system theory and theory of open systems, following the line of the present writer, is given in W. Beier's *Biophysik* (expected soon to be available in English). An axiomatic development is G. J. Klir's *An Approach to General Systems Theory*. For system theory developed from the viewpoint of control technology, H. Schwarz, *Einführung in die moderne Systemtheorie* is suggested. For system theory in the sciences of man, the following are important: *General Systems Theory and Psychiatry* (W. Gray, F. D. Duhl and N. D. Rizzo, eds.); *Modern Systems Research for the Behavioral Scientist* (W. Buckley, ed.);

System, Change and Conflict (N. J. Demerath and R. A. Peterson, eds.). System philosophy is developed in Laszlo's *Introduction to Systems Philosophy.*

While the text of the original edition is unchanged (except for corrections of some misprints), the preface to this new edition, the appendix, "Notes on Developments in Mathematical System Theory," and the "Addenda" to Suggestions for Further Readings were added. It is hoped that this book will further serve as an introduction to students, and a stimulus to workers in general system theory.

General System Theory

Foundations, Development, Applications

1 Introduction

Systems Everywhere

If someone were to analyze current notions and fashionable catchwords, he would find "systems" high on the list. The concept has pervaded all fields of science and penetrated into popular thinking, jargon and mass media. Systems thinking plays a dominant role in a wide range of fields from industrial enterprise and armaments to esoteric topics of pure science. Innumerable publications, conferences, symposia and courses are devoted to it. Professions and jobs have appeared in recent years which, unknown a short while ago, go under names such as systems design, systems analysis, systems engineering and others. They are the very nucleus of a new technology and technocracy; their practitioners are the "new utopians" of our time (Boguslaw, 1965) who —in contrast to the classic breed whose ideas remained between the covers of books—are at work creating a New World, brave or otherwise.

The roots of this development are complex. One aspect is the development from power engineering—that is, release of large amounts of energy as in steam or electric machines—to control engineering, which directs processes by low-power devices and has led to computers and automation. Self-controlling machines have appeared, from the humble domestic thermostat to the self-steering missiles of World War II to the immensely improved missiles of today. Technology has been led to think not in terms of single machines but in those of "systems." A steam engine, automobile, or radio receiver was within the competence of the engineer trained in the respective specialty. But when it comes to ballistic missiles or space vehicles, they have to be assembled from components originating in heterogeneous technologies, mechanical, electronic, chemical, etc.; relations of man and machine come into play; and innumerable financial, economic, social and political problems are thrown into the bargain. Again, air or even automobile traffic are not just a matter of the number of vehicles in operation, but are systems to be planned or arranged. So innumerable problems are arising in production, commerce, and armaments.

Thus, a "systems approach" became necessary. A certain objective is given; to find ways and means for its realization requires the systems specialist (or team of specialists) to consider alternative solutions and to choose those promising optimization at maximum efficiency and minimal cost in a tremendously complex network of interactions. This requires elaborate techniques and computers for solving problems far transcending the capacity of an individual mathematician. Both the "hardware" of computers, automation and cybernation, and the "software" of systems science represent a new technology. It has been called the Second Industrial Revolution and has developed only in the past few decades.

These developments have not been limited to the industrial-military complex. Politicians frequently ask for application of the "systems approach" to pressing problems such as air and water pollution, traffic congestion, urban blight, juvenile delinquency and organized crime, city planning (Wolfe, 1967), etc., designating this a "revolutionary new concept" (Carter, 1966; Boffey, 1967). A Canadian Premier (Manning, 1967) writes the systems approach into his political platform saying that

> an interrelationship exists between all elements and constituents of society. The essential factors in public problems, issues, policies, and programs must always be considered and evaluated as interdependent components of a total system.

These developments would be merely another of the many facets of change in our contemporary technological society were it not for a significant factor apt to be overlooked in the highly sophisticated and necessarily specialized techniques of computer science, systems engineering and related fields. This is not only a tendency in technology to make things bigger and better (or alternatively, more profitable, destructive, or both). It is a change in basic categories of thought of which the complexities of modern technology are only one—and possibly not the most important—manifestation. In one way or another, we are forced to deal with complexities, with "wholes" or "systems," in all fields of knowledge. This implies a basic re-orientation in scientific thinking.

An attempt to summarize the impact of "systems" would not be feasible and would pre-empt the considerations of this book. A few examples, more or less arbitrarily chosen, must suffice to outline the nature of the problem and consequent re-orientation. The reader should excuse an egocentric touch in the quotations, in view of the fact that the purpose of this book is to present the author's viewpoint rather than a neutral review of the field.

In physics, it is well-known that in the enormous strides it made in the past few decades, it also generated new problems—or possibly a new kind of problem—perhaps most evident to the laymen in the indefinite number of some hundreds of elementary particles for which at present physics can offer little rhyme or reason. In the words of a noted representative (de-Shalit, 1966), the further development of nuclear physics "requires much experimental work, as well as the development of additional powerful methods for the handling of systems with many, but not infinitely many, particles." The same quest was expressed by A. Szent-Györgyi (1964), the great physiologist, in a whimsical way:

> [When I joined the Institute for Advanced Study in Princeton] I did this in the hope that by rubbing elbows with those great atomic physicists and mathematicians I would learn something about living matters. But as soon as I revealed that in any living system there are more than two electrons, the physicists would not speak to me. With all their computers they could not say what the third electron might do. The remarkable thing is that it knows exactly what to do. So that little electron knows something that all the wise men of Princeton don't, and this can only be something very simple.

And Bernal (1957), some years ago, formulated the still-unsolved problems thus:

> No one who knows what the difficulties are now believes that the crisis of physics is likely to be resolved by any simple trick

or modification of existing theories. Something radical is needed, and it will have to go far wider than physics. A new world outlook is being forged, but much experience and argument will be needed before it can take a definitive form. It must be coherent, it must include and illuminate the new knowledge of fundamental particles and their complex fields, it must resolve the paradoxes of wave and particle, it must make the world inside the atom and the wide spaces of the universe equally intelligible. It must have a different dimension from all previous world views, and include in itself an explanation of development and the origin of new things. In this it will fall naturally in line with the converging tendencies of the biological and social sciences in which a regular pattern blends with their evolutionary history.

The triumph in recent years of molecular biology, the "breaking" of the genetic code, the consequent achievements in genetics, evolution, medicine, cell physiology and many other fields, has become common knowledge. But in spite of—or just because of—the deepened insight attained by "molecular" biology, the necessity of "organismic" biology has become apparent, as this writer had advocated for some 40 years. The concern of biology is not only at the physico-chemical or molecular level but at the higher levels of living organization as well. As we shall discuss later (p. 12), the demand has been posed with renewed strength in consideration of recent facts and knowledge; but hardly an argument not previously discussed (von Bertalanffy, 1928a, 1932, 1949a, 1960) has been added.

Again, the basic conception in psychology used to be the "robot model." Behavior was to be explained by the mechanistic stimulus-response (S-R) scheme; conditioning, according to the pattern of animal experiment, appeared as the foundation of human behavior; "meaning" was to be replaced by conditioned response; specificity of human behavior to be denied, etc. Gestalt psychology first made an inroad into the mechanistic scheme some 50 years ago. More recently, many attempts toward a more satisfactory "image of man" have been made and the system concept is gaining in importance (Chapter 8); Piaget, for example, "expressly related his conceptions to the general system theory of Bertalanffy" (Hahn, 1967).

Perhaps even more than psychology, psychiatry has taken up the systems viewpoint (e.g. Menninger, 1963; von Bertalanffy, 1966; Grinker, 1967; Gray et al., 1969). To quote from Grinker:

> Of the so-called global theories the one initially stated and defined by Bertalanffy in 1947 under the title of "general systems theory" has taken hold. . . . Since then he has refined, modified and applied his concepts, established a society for general systems theory and published a *General Systems Yearbook*. Many social scientists but only a handful of psychiatrists studied, understood or applied systems theory. Suddenly, under the leadership of Dr. William Gray of Boston, a threshold was reached so that at the 122nd annual meeting of the American Psychiatric Association in 1966 two sessions were held at which this theory was discussed and regular meetings for psychiatrists were ensured for future participation in a development of this "Unified Theory of Human Behavior." If there be a third revolution (i.e. after the psychoanalytic and behavioristic), it is in the development of a general [system] theory (p. ix).

A report of a recent meeting (American Psychiatric Association, 1967) draws a vivid picture:

> When a room holding 1,500 people is so jammed that hundreds stand through an entire morning session, the subject must be one in which the audience is keenly interested. This was the situation at the symposium on the use of a general systems theory in psychiatry which took place at the Detroit meeting of the American Psychiatric Association (Damude, 1967).

The same in the social sciences. From the broad spectrum, widespread confusion and contradictions of contemporary sociological theories (Sorokin, 1928, 1966), one secure conclusion emerges: that social phenomena must be considered as "systems" —difficult and at present unsettled as the definition of sociocultural entities may be. There is

> a revolutionary scientific perspective (stemming) from the General Systems Research movement and (with a) wealth of principles, ideas and insights that have already brought a higher degree of scientific order and understanding to many

areas of biology, psychology and some physical sciences.... Modern systems research can provide the basis of a framework more capable of doing justice to the complexities and dynamic properties of the socio-cultural system (Buckley, 1967).

The course of events in our times suggests a similar conception in history, including the consideration that, after all, history is sociology in the making or in "longitudinal" study. It is the same socio-cultural entities which sociology investigates in their present state and history in their becoming.

Earlier periods of history may have consoled themselves by blaming atrocities and stupidities on bad kings, wicked dictators, ignorance, superstition, material want and related factors. Consequently, history was of the "who-did-what" kind—"idiographic," as it was technically known. Thus the Thirty-Years War was a consequence of religious superstition and the rivalries of German princes; Napoleon overturned Europe because of his unbridled ambition; the Second World War could be blamed on the wickedness of Hitler and the warlike proclivity of the Germans.

We have lost this intellectual comfort. In a state of democracy, universal education and general affluence, these previous excuses for human atrocity fail miserably. Contemplating contemporary history in the making, it is difficult to ascribe its irrationality and bestiality solely to individuals (unless we grant them a super-human—or subhuman—capacity for malice and stupidity). Rather, we seem to be victims of "historical forces"—whatever this may mean. Events seem to involve more than just individual decisions and actions and to be determined more by socio-cultural "systems," be these prejudices, ideologies, pressure groups, social trends, growth and decay of civilizations, or what not. We know precisely and scientifically what the effects of pollution, waste of natural resources, the population explosion, the armaments race, etc., are going to be. We are told so every day by countless critics citing irrefutable arguments. But neither national leaders nor society as a whole seems to be able to do anything about it. If we do not want a theistic explanation—*Quem Deus perdere vult dementat*—we seem to follow some tragic historical necessity.

While realizing the vagueness of such concepts as civilization and the shortcomings of "grand theories" like those of Spengler and Toynbee, the question of regularities or laws of socio-cultural systems makes sense though this does not necessarily mean historical inevitability according to Sir Isaiah Berlin. An historical panorama like McNeill's *The Rise of the West* (1963), which indicates his anti-Spenglerian position even in the title, nevertheless is a story of historical systems. Such a conception penetrates into seemingly outlying fields so that the view of the "process-school of archaeology" is said to be "borrowed from Ludwig von Bertalanffy's framework for the developing embryo, where systems trigger behavior at critical junctures and, once they have done so, cannot return to their original pattern" (Flannery, 1967).

While sociology (and presumably history) deals with informal organizations, another recent development is the theory of formal organizations, that is, structures planfully instituted, such as those of an army, bureaucracy, business enterprise, etc. This theory is "framed in a philosophy which accepts the premise that the only meaningful way to study organization is to study it as a system," systems analysis treating "organization as a system of mutually dependent variables"; therefore "modern organization theory leads almost inevitably into a discussion of general system theory" (Scott, 1963). In the words of a practitioner of operational research,

> In the last two decades we have witnessed the emergence of the "system" as a key concept in scientific research. Systems, of course, have been studied for centuries, but something new has been added.... The tendency to study systems as an entity rather than as a conglomeration of parts is consistent with the tendency in contemporary science no longer to isolate phenomena in narrowly confined contexts, but rather to open interactions for examination and to examine larger and larger slices of nature. Under the banner of *systems research* (and its many synonyms) we have also witnessed a convergence of many more specialized contemporary scientific developments. ... These research pursuits and many others are being interwoven into a cooperative research effort involving an everwidening spectrum of scientific and engineering disciplines. We are participating in what is probably the most comprehensive effort to attain a synthesis of scientific knowledge yet made (Ackoff, 1959).

In this way, the circle closes and we come back to those developments in contemporary technological society with which we started. What emerges from these considerations—however sketchy and superficial—is that in the gamut of modern sciences and life new conceptualizations, new ideas and categories are required, and that these, in one way or another, are centered about the concept of "system." To quote, for a change, from a Soviet author:

> The elaboration of specific methods for the investigation of systems is a general trend of present scientific knowledge, just as for 19th century science the primary concentration of attention to the elaboration of elementary forms and processes in nature was characteristic (Lewada, in Hahn, 1967, p. 185).

The dangers of this new development, alas, are obvious and have often been stated. The new cybernetic world, according to the psychotherapist Ruesch (1967) is not concerned with people but with "systems"; man becomes replaceable and expendable. To the new utopians of systems engineering, to use a phrase of Boguslaw (1965), it is the "human element" which is precisely the unreliable component of their creations. It either has to be eliminated altogether and replaced by the hardware of computers, self-regulating machinery and the like, or it has to be made as reliable as possible, that is, mechanized, conformist, controlled and standardized. In somewhat harsher terms, man in the Big System is to be—and to a large extent has become—a moron, button-pusher or learned idiot, that is, highly trained in some narrow specialization but otherwise a mere part of the machine. This conforms to a well-known systems principle, that of progressive mechanization—the individual becoming ever more a cogwheel dominated by a few privileged leaders, mediocrities and mystifiers who pursue their private interests under a smokescreen of ideologies (Sorokin, 1966, pp. 558ff).

Whether we envisage the positive expansion of knowledge and beneficent control of environment and society, or see in the systems movement the arrival of *Brave New World* and *1984*—it deserves intensive study, and we have to come to terms with it.

On the History of Systems Theory

As we have seen, there is a consensus in all major fields— from subatomic physics to history—that a re-orientation of science is due. Developments in modern technology parallel this trend.

So far as can be ascertained, the idea of a "general system

theory" was first introduced by the present author prior to cybernetics, systems engineering and the emergence of related fields. The story of how he was led to this notion is briefly told elsewhere in this book (pp. 89ff.), but some amplification appears to be in order in view of recent discussions.

As with every new idea in science and elsewhere, the systems concept has a long history. Although the term "system" itself was not emphasized, the history of this concept includes many illustrious names. As "natural philosophy," we may trace it back to Leibniz; to Nicholas of Cusa with his coincidence of opposites; to the mystic medicine of Paracelsus; to Vico's and ibn-Khaldun's vision of history as a sequence of cultural entities or "systems"; to the dialectic of Marx and Hegel, to mention but a few names from a rich panoply of thinkers. The literary gourmet may remember Nicholas of Cusa's *De ludo globi* (1463; cf. von Bertalanffy, 1928b) and Hermann Hesse's *Glasperlenspiel*, both of them seeing the working of the world reflected in a cleverly designed, abstract game.

There had been a few preliminary works in the field of general system theory. Köhler's "physical *gestalten*" (1924) pointed in this direction but did not deal with the problem in full generality, restricting its treatment to *gestalten* in physics (and biological and psychological phenomena presumably interpretable on this basis). In a later publication (1927), Köhler raised the postulate of a system theory, intended to elaborate the most general properties of inorganic compared to organic systems; to a degree, this demand was met by the theory of open systems. Lotka's classic (1925) came closest to the objective, and we are indebted to him for basic formulations. Lotka indeed dealt with a general concept of systems (not, like Köhler's, restricted to systems of physics). Being a statistician, however, with his interest lying in population problems rather than in biological problems of the individual organism, Lotka, somewhat strangely, conceived communities as systems, while regarding the individual organism as a sum of cells.

Nevertheless, the necessity and feasibility of a systems approach became apparent only recently. Its necessity resulted from the fact that the mechanistic scheme of isolable causal trains and meristic treatment had proved insufficient to deal with theoretical problems, especially in the biosocial sciences, and with the practical problems posed by modern technology. Its feasibility resulted from various new developments—theoretical, epistemological, mathematical, etc.—which, although still in their beginnings, made it progressively realizable.

The present author, in the early 20's, became puzzled about obvious lacunae in the research and theory of biology. The then prevalent mechanistic approach just mentioned appeared to neglect or actively deny just what is essential in the phenomena of life. He advocated an organismic conception in biology which emphasizes consideration of the organism as a whole or system, and sees the main objective of biological sciences in the discovery of the principles of organization at its various levels. The author's first statements go back to 1925–26, while Whitehead's philosophy of "organic mechanism" was published in 1925. Cannon's work on homeostasis appeared in 1929 and 1932. The organismic conception had its great precursor in Claude Bernard, but his work was hardly known outside France; even now it awaits its full evaluation (cf. Bernal, 1957, p. 960). The simultaneous appearance of similar ideas independently and on different continents was symptomatic of a new trend which, however, needed time to become accepted.

These remarks are prompted by the fact that in recent years "organismic biology" has been re-emphasized by leading American biologists (Dubos, 1964, 1967; Dobzhansky, 1966; Commoner, 1961) without, however, mentioning the writer's much earlier work, although this is duly recognized in the literature of Europe and of the socialist countries (e.g., Ungerer, 1966; Blandino, 1960; Tribiño, 1946; Kanaev, 1966; Kamarýt, 1961, 1963; Bendmann, 1963, 1967; Afanasjew, 1962). It can be definitely stated that recent discussions (e.g., Nagel, 1961; Hempel, 1965; Beckner, 1959; Smith, 1966; Schaffner, 1967), although naturally referring to advances of biology in the past 40 years, have not added any new viewpoints in comparison to the author's work.

In philosophy, the writer's education was in the tradition of neopositivism of the group of Moritz Schlick which later became known as the Vienna Circle. Obviously, however, his interest in German mysticism, the historical relativism of Spengler and the history of art, and similar unorthodox attitudes precluded his becoming a good positivist. Stronger were his bonds with the Berlin group of the "Society for Empirical Philosophy" of the

1920's, in which the philosopher-physicist Hans Reichenbach, the psychologist A. Herzberg, the engineer Parseval (inventor of dirigible aircraft) were prominent.

In connection with experimental work on metabolism and growth on the one hand, and an effort to concretize the organismic program on the other, the theory of open systems was advanced, based on the rather trivial fact that the organism happens to be an open system, but no theory existed at the time. The first presentation, which followed some tentative trials, is included in this volume (Chapter 5). Biophysics thus appeared to demand an expansion of conventional physical theory in the way of generalization of kinetic principles and thermodynamic theory, the latter becoming known, later on, as irreversible thermodynamics.

But then, a further generalization became apparent. In many phenomena in biology and also in the behavioral and social sciences, mathematical expressions and models are applicable. These, obviously, do not pertain to the entities of physics and chemistry, and in this sense transcend physics as the paragon of "exact science." (Incidentally, a series *Abhandlungen zur exakten Biologie*, in succession of Schaxel's previous *Abhandlungen zur theoretischen Biologie*, was inaugurated by the writer but stopped during the war.) The structural similarity of such models and their isomorphism in different fields became apparent; and just those problems of order, organization, wholeness, teleology, etc., appeared central which were programmatically excluded in mechanistic science. This, then, was the idea of "general system theory."

The time was not favorable for such development. Biology was understood to be identical with laboratory work, and the writer had already gone out on a limb when publishing *Theoretische Biologie* (1932), another field which has only recently become academically respectable. Nowadays, when there are numerous journals and publications in this discipline and model building has become a fashionable and generously supported indoor sport, the resistance to such ideas is hard to imagine. Affirmation of the concept of general system theory, especially by the late Professor Otto Pötzl, well-known Vienna psychiatrist, helped the writer to overcome his inhibitions and to issue a statement (reproduced in Chapter 3 of this book). Again, fate intervened. The paper (in *Deutsche Zeitschrift für Philosophie*) had reached the proof stage, but the issue to carry it was destroyed in the catastrophe of the last war. After the war, general system theory was presented in lectures (cf. Appendix), amply discussed with physicists (von Bertalanffy, 1948a) and discussed in lectures and symposia (e.g., von Bertalanffy et al., 1951).

The proposal of system theory was received incredulously as fantastic or presumptuous. Either—it was argued—it was *trivial* because the so-called isomorphisms were merely examples of the truism that mathematics can be applied to all sorts of things, and it therefore carried no more weight than the "discovery" that $2 + 2 = 4$ holds true for apples, dollars and galaxies alike; or it was *false* and *misleading* because superficial analogies—as in the famous simile of society as an "organism"—camouflage actual differences and so lead to wrong and even morally objectionable conclusions. Or, again, it was philosophically and methodologically *unsound* because the alleged "irreducibility" of higher levels to lower ones tended to impede analytical research whose success was obvious in various fields such as in the reduction of chemistry to physical principles, or of life phenomena to molecular biology.

Gradually it was realized that such objections missed the point of what systems theory stands for, namely, attempting scientific interpretation and theory where previously there was none, and higher generality than that in the special sciences. General system theory responded to a secret trend in various disciplines. A letter from K. Boulding, economist, dated 1953, well summarized the situation:

> I seem to have come to much the same conclusion as you have reached, though approaching it from the direction of economics and the social sciences rather than from biology— that there is a body of what I have been calling "general empirical theory," or "general system theory" in your excellent terminology, which is of wide applicability in many different disciplines. I am sure there are many people all over the world who have come to essentially the same position that we have, but we are widely scattered and do not know each other, so difficult is it to cross the boundaries of the disciplines.

In the first year of the Center for Advanced Study in the Behavioral Sciences (Palo Alto), Boulding, the biomathematician

A. Rapoport, the physiologist Ralph Gerard and the present writer found themselves together. The project of a Society for General System Theory was realized at the Annual Meeting of the American Association for the Advancement of Science in 1954. The name was later changed into the less pretentious "Society for General Systems Research," which is now an affiliate of the AAAS and whose meetings have become a well-attended fixture of the AAAS conventions. Local groups of the Society were established at various centers in the United States and subsequently in Europe. The original program of the Society needed no revision:

> The Society for General Systems Research was organized in 1954 to further the development of theoretical systems which are applicable to more than one of the traditional departments of knowledge. Major functions are to: (1) investigate the isomorphy of concepts, laws, and models in various fields, and to help in useful transfers from one field to another; (2) encourage the development of adequate theoretical models in the fields which lack them; (3) minimize the duplication of theoretical effort in different fields; (4) promote the unity of science through improving communication among specialists.

The Society's Yearbooks, *General Systems*, under the efficient editorship of A. Rapoport, have since served as its organ. Intentionally *General Systems* does not follow a rigid policy but rather provides a place for working papers of different intention as seems to be appropriate in a field which needs ideas and exploration. A large number of investigations and publications substantiated the trend in various fields; a journal, *Mathematical Systems Theory*, made its appearance.

Meanwhile another development had taken place. Norbert Wiener's *Cybernetics* appeared in 1948, resulting from the then recent developments of computer technology, information theory, and self-regulating machines. It was again one of the coincidences occurring when ideas are in the air that three fundamental contributions appeared at about the same time: Wiener's *Cybernetics* (1948), Shannon and Weaver's information theory (1949) and von Neumann and Morgenstern's game theory (1947). Wiener carried the cybernetic, feedback and information concepts far beyond the fields of technology and generalized it in the biological and social realms. It is true that cybernetics was not without precursors. Cannon's concept of homeostasis became a cornerstone in these considerations. Less well-known, detailed feedback models of physiological phenomena had been elaborated by the German physiologist Richard Wagner (1954) in the 1920's, the Swiss Nobel prize winner W. R. Hess (1941, 1942) and in Erich von Holst's *Reafferenzprinzip*. The enormous popularity of cybernetics in science, technology and general publicity is, of course, due to Wiener and his proclamation of the Second Industrial Revolution.

The close correspondence of the two movements is well shown in a programmatic statement of L. Frank introducing a cybernetics conference:

The concepts of purposive behavior and teleology have long been associated with a mysterious, self-perfecting or goal-seeking capacity or final cause, usually of superhuman or super-natural origin. To move forward to the study of events, scientific thinking had to reject these beliefs in purpose and these concepts of teleological operations for a strictly mechanistic and deterministic view of nature. This mechanistic conception became firmly established with the demonstration that the universe was based on the operation of anonymous particles moving at random, in a disorderly fashion, giving rise, by their multiplicity, to order and regularity of a statistical nature, as in classical physics and gas laws. The unchallenged success of these concepts and methods in physics and astronomy, and later in chemistry, gave biology and physiology their major orientation. This approach to problems of organisms was reinforced by the analytical preoccupation of the Western European culture and languages. The basic assumptions of our traditions and the persistent implications of the language we use almost compel us to approach everything we study as composed of separate, discrete parts or factors which we must try to isolate and identify as potent causes. Hence, we derive our preoccupation with the study of the relation of two variables. We are witnessing today a search for new approaches, for new and more comprehensive concepts and for methods capable of dealing with the large wholes of organisms and personalities. The concept of teleological mechanisms, however it may be expressed in different terms, may be viewed as an attempt to escape from these older mechanistic formulations that now appear inadequate, and to provide new and more fruitful conceptions and more effective methodologies for studying self-regulating processes, self-orientating systems and organisms, and self-directing personalities. Thus, the terms *feedback, servo-mechanisms, circular systems,* and *circular processes* may be viewed as different but equivalent expressions of much the same basic conception. (Frank *et al.,* 1948, condensed).

A review of the development of cybernetics in technology and science would exceed the scope of this book, and is unnecessary in view of the extensive literature of the field. However, the present historical survey is appropriate because certain misunderstandings and misinterpretations have appeared. Thus Buckley (1967, p. 36) states that "modern Systems Theory, though seemingly springing de novo out of the last war effort, can be seen as a culmination of a broad shift in scientific perspective striving for dominance over the last few centuries." Although the second part of the sentence is true, the first is not; systems theory did not "spring out of the last war effort," but goes back much further and had roots quite different from military hardware and related technological developments. Neither is there an "emergence of system theory from recent developments in the analysis of engineering systems" (Shaw, 1965) except in a special sense of the word.

Systems theory also is frequently identified with cybernetics and control theory. This again is incorrect. Cybernetics, as the theory of control mechanisms in technology and nature and founded on the concepts of information and feedback, is but a part of a general theory of systems; cybernetic systems are a special case, however important, of systems showing self-regulation.

Trends in Systems Theory

At a time when any novelty, however trivial, is hailed as being revolutionary, one is weary of using this label for scientific developments. Miniskirts and long hair being called teenage revolution, and any new styling of automobiles or drug introduced by the pharmaceutical industry being so announced, the word is an advertising slogan hardly fit for serious consideration. It can, however, be used in a strictly technical sense, i.e., "scientific revolutions" can be identified by certain diagnostic criteria.

Following Kuhn (1962), a scientific revolution is defined by the appearance of new conceptual schemes or "paradigms." These bring to the fore aspects which previously were not seen or perceived, or even suppressed in "normal" science, i.e., science generally accepted and practiced at the time. Hence there is a shift in the problems noticed and investigated and a change of the rules of scientific practice, comparable to the switch in perceptual gestalten in psychological experiments, when, e.g., the same figure may be seen as two faces vs. cup, or as duck vs. rabbit. Understandably, in such critical phases emphasis is laid on philosophical analysis which is not felt necessary in periods of growth of "normal" science. The early versions of a new paradigm are mostly crude, solve few problems, and solutions given for individual problems are far from perfect. There is a profusion and competition of theories, each limited with respect to the number of problems covered, and elegant solution of those taken into account. Nevertheless, the new paradigm does cover new problems, especially those previously rejected as "metaphysical".

These criteria were derived by Kuhn from a study of the "classical" revolutions in physics and chemistry, but they are an excellent description of the changes brought about by organismic and systems concepts, and elucidate both their merits and limitations. Especially and not surprisingly, systems theory comprises a number of approaches different in style and aims.

The system problem is essentially the problem of the limitations of analytical procedures in science. This used to be expressed by half-metaphysical statements, such as emergent evolution or "the whole is more than a sum of its parts," but has a clear operational meaning. "Analytical procedure" means that an entity investigated be resolved into, and hence can be constituted or reconstituted from, the parts put together, these procedures being understood both in their material and conceptual sense. This is the basic principle of "classical" science, which can be circumscribed in different ways: resolution into isolable causal trains, seeking for "atomic" units in the various fields of science, etc. The progress of science has shown that these principles of classical science—first enunciated by Galileo and Descartes—are highly successful in a wide realm of phenomena.

Application of the analytical procedure depends on two conditions. The first is that interactions between "parts" be nonexistent or weak enough to be neglected for certain research purposes. Only under this condition, can the parts be "worked out," actually, logically, and mathematically, and then be "put together." The second condition is that the relations describing the behavior of parts be linear; only then is the condition of summativity given, i.e., an equation describing the behavior of the total is of the same form as the equations describing the behavior of the parts; partial processes can be superimposed to obtain the total process, etc.

These conditions are not fulfilled in the entities called systems, i.e., consisting of parts "in interaction." The prototype of their description is a set of simultaneous differential equations (pp. 55ff.), which are nonlinear in the general case. A system or "organized complexity" (p. 34) may be circumscribed by the existence of "strong interactions" (Rapoport, 1966) or interactions which are "nontrivial" (Simon, 1965), i.e., nonlinear. The methodological problem of systems theory, therefore, is to provide for problems which, compared with the analytical-summative ones of classical science, are of a more general nature.

As has been said, there are various approaches to deal with such problems. We intentionally use the somewhat loose expression "approaches" because they are logically inhomogeneous, represent different conceptual models, mathematical techniques, general points of view, etc.; they are, however, in accord in being "systems theories." Leaving aside approaches in applied systems research, such as systems engineering, operational research, linear and nonlinear programming, etc., the more important approaches are as follows. (For a good survey, cf. Drischel, 1968).

"Classical" system theory applies classical mathematics, i.e., calculus. Its aim is to state principles which apply to systems in general or defined subclasses (e.g., closed and open systems), to provide techniques for their investigation and description, and to apply these to concrete cases. Owing to the generality of such description, it may be stated that certain formal properties will apply to any entity *qua* system (or open system, or hierarchical

system, etc.), even when its particular nature, parts, relations, etc., are unknown or not investigated. Examples include generalized principles of kinetics applicable, e.g., to populations of molecules or biological entities, i.e., to chemical and ecological systems; diffusion, such as diffusion equations in physical chemistry and in the spread of rumors; application of steady state and statistical mechanics models to traffic flow (Gazis, 1967); allometric analysis of biological and social systems.

Computerization and simulation. Sets of simultaneous differential equations as a way to "model" or define a system are, if linear, tiresome to solve even in the case of a few variables; if nonlinear, they are unsolvable except in special cases (Table 1.1).

Table 1.1

Classification of Mathematical Problems* and Their Ease of Solution by Analytical Methods. After Franks, 1967.

	Linear Equations			Nonlinear Equations		
Equation	One Equation	Several Equations	Many Equations	One Equation	Several Equations	Many Equations
Algebraic	Trivial	Easy	Essentially impossible	Very difficult	Very difficult	Impossible
Ordinary differential	Easy	Difficult	Essentially impossible	Very difficult	Impossible	Impossible
Partial differential	Difficult	Essentially impossible	Impossible	Impossible	Impossible	Impossible

* Courtesy of Electronic Associates, Inc.

For this reason, computers have opened a new approach in systems research; not only by way of facilitation of calculations which otherwise would exceed available time and energy and by replacement of mathematical ingenuity by routine procedures, but also by opening up fields where no mathematical theory or ways of solution exist. Thus systems far exceeding conventional mathematics can be computerized; on the other hand, actual laboratory experiment can be replaced by computer simulation, the model so developed then to be checked by experimental data. In such way, for example, B. Hess has calculated the fourteen-step reaction chain of glycolysis in the cell in a model of more than 100 nonlinear differential equations. Similar analyses are routine in economics, market research, etc.

Compartment theory. An aspect of systems which may be listed separately because of the high sophistication reached in the field is compartment theory (Rescigno and Segre, 1966), i.e., the system consists of subunits with certain boundary conditions between which transport processes take place. Such compartment systems may have, e.g., "catenary" or "mammillary" structure (chain of compartments or a central compartment communicating with a number of peripheral ones). Understandably, mathematical difficulties become prohibitive in the case of three- or multicompartment systems. Laplace transforms and introduction of net and graph theory make analysis possible.

Set theory. The general formal properties of systems, closed and open systems, etc., can be axiomatized in terms of set theory (Mesarović, 1964; Maccia, 1966). In mathematical elegance this approach compares favorably with the cruder and more special formulations of "classical" system theory. The connections of axiomatized systems theory (or its present beginnings) with actual systems problems are somewhat tenuous.

Graph theory. Many systems problems concern structural or topologic properties of systems, rather than quantitative relations. Some approaches are available in this respect. Graph theory, especially the theory of directed graphs (digraphs), elaborates relational structures by representing them in a topological space. It has been applied to relational aspects of biology (Rashevsky, 1956, 1960; Rosen, 1960). Mathematically, it is connected with matrix algebra; modelwise, with compartment theory of systems containing partly "permeable" subsystems, and from here with the theory of open systems.

Net theory, in its turn, is connected with set, graph, compartment, etc., theories and is applied to such systems as nervous networks (e.g., Rapoport, 1949–50).

Cybernetics is a theory of control systems based on communication (transfer of information) between system and environment and within the system, and control (feedback) of the system's function in regard to environment. As mentioned and to be discussed further, the model is of wide application but should not be identified with "systems theory" in general. In biology and other basic sciences, the cybernetic model is apt to describe the formal structure of regulatory mechanisms, e.g., by block and flow diagrams. Thus the regulatory structure can be recognized, even when actual mechanisms remain unknown and undescribed, and the system is a "black box" defined only by input and output. For similar reasons, the same cybernetic scheme may apply to hydraulic, electric, physiological, etc., systems. The highly elaborate and sophisticated theory of servomechanism in technology has been applied to natural systems only in a limited extent (cf. Bayliss, 1966; Kalmus, 1966; Milsum, 1966).

Information theory, in the sense of Shannon and Weaver (1949), is based on the concept of information, defined by an expression isomorphic to negative entropy of thermodynamics. Hence the expectation that information may be used as measure of organization (cf. p. 42; Quastler, 1955). While information theory gained importance in communication engineering, its applications to science have remained rather unconvincing (E.N. Gilbert, 1966). The relationship between information and organization, information theory and thermodynamics, remains a major problem (cf. pp. 151ff.).

Theory of automata (see Minsky, 1967) is the theory of abstract automata, with input, output, possibly trial-and-error and learning. A general model is the Turing machine (1936). Expressed in the simplest way a Turing automaton is an abstract machine capable of imprinting (or deleting) "1" and "0" marks on a tape of infinite length. It can be shown that any process of whatever complexity can be simulated by a machine, if this process can be expressed in a finite number of logical operations. Whatever is possible logically (i.e., in an algorithmic symbolism) also can be construed—in principle, though of course by no means always in practice—by an automaton, i.e., an algorithmic machine.

Game theory (von Neumann and Morgenstern, 1947) is a different approach but may be ranged among systems sciences because it is concerned with the behavior of supposedly "rational" players to obtain maximal gains and minimal losses by appropriate strategies against the other player (or nature). Hence it concerns essentially a "system" of antagonistic "forces" with specifications.

Decision theory is a mathematical theory concerned with choices among alternatives.

Queuing theory concerns optimization of arrangements under conditions of crowding.

Inhomogeneous and incomplete as it is, confounding models (e.g., open system, feedback circuit) with mathematical techniques (e.g., set, graph, game theory), such an enumeration is apt to show that there is an array of approaches to investigate systems, including powerful mathematical methods. The point to be reiterated is that problems previously not envisaged, not manageable, or considered as being beyond science or purely philosophical are progressively explored.

Naturally, an incongruence between model and reality often exists. There are highly elaborate and sophisticated mathematical models, but it remains dubious how they can be applied to the concrete case; there are fundamental problems for which no mathematical techniques are available. Disappointment of over-extended expectations has occurred. Cybernetics, e.g., proved its impact not only in technology but in basic sciences, yielding models for concrete phenomena and bringing teleological phenomena—previously tabooed—into the range of scientifically legitimate problems; but it did not yield an all-embracing explanation or grand "world view," being an extension rather than a replacement of the mechanistic view and machine theory (cf. Bronowski, 1964). Information theory, highly developed mathematically, proved disappointing in psychology and sociology. Game theory was hopefully applied to war and politics; but one hardly feels that it has led to an improvement of political decisions and the state of the world; a failure not unexpected when considering how little the powers that be resemble the "rational" players of game theory. Concepts and models of equilibrium, homeostasis, adjustment, etc., are suitable for the maintenance of systems, but inadequate for phenomena of change, differentiation, evolution, negentropy, production of improbable states, creativity, building-up of tensions, self-realization, emergence, etc.; as indeed Cannon realized when he acknowledged, beside homeostasis, a "heterostasis" including phenomena of the latter nature. The theory of open systems applies to a wide range of phenomena in biology (and technology), but a warning is necessary against its incautious expansion to fields for which its concepts are not made. Such limitations and lacunae are only what is to be expected in a field hardly older than twenty or thirty years. In the last resort, disappointment results from making what is a useful model in certain respects into some metaphysical reality and "nothing-but" philosophy, as has happened many times in intellectual history.

The advantages of mathematical models—unambiguity, possibility of strict deduction, verifiability by observed data—are well known. This does not mean that models formulated in ordinary language are to be despised or refused.

A *verbal model* is better than no model at all, or a model which, because it can be formulated mathematically, is forcibly imposed upon and falsifies reality. Theories of enormous influence such as psychoanalysis were unmathematical or, like the theory of selection, their impact far exceeded mathematical constructions which came only later and cover only partial aspects and a small fraction of empirical data.

Mathematics essentially means the existence of an algorithm which is much more precise than that of ordinary language. History of science attests that expression in ordinary language often preceded mathematical formulation, i.e., invention of an algorithm. Examples come easily to mind: the evolution from counting in words to Roman numerals (a semiverbal, clumsy, half-algorithm) to Arabic notation with position value; equations, from verbal formulation to rudimentary symbolism handled with virtuosity (but difficult for us to follow) by Diophantus and other founders of algebra, to modern notation; theories like those of Darwin or of economics which only later found a (partial) mathematical formulation. It may be preferable first to have some nonmathematical model with its shortcomings but expressing some previously unnoticed aspect, hoping for future development of a suitable algorithm, than to start with premature mathematical models following known algorithms and, therefore, possibly restricting the field of vision. Many developments in molecular biology, theory of selection, cybernetics and other fields showed the blinding effects of what Kuhn calls "normal" science, i.e., monolithically accepted conceptual schemes.

Models in ordinary language therefore have their place in systems theory. The system idea retains its value even where it cannot be formulated mathematically, or remains a "guiding idea" rather than being a mathematical construct. For example, we may not have satisfactory system concepts in sociology; the mere insight that social entities are systems rather than sums of social atoms, or that history consists of systems (however ill defined) called civilizations obeying principles general to systems, implies a reorientation in the fields concerned.

As can be seen from the above survey, there are, within the "systems approach," mechanistic and organismic trends and models, trying to master systems either by "analysis," "linear (including circular) causality," "automata," or else by "whole-ness," "interaction," "dynamics" (or what other words may be used to circumscribe the difference). While these models are not mutually exclusive and the same phenomena may even be approached by different models (e.g., "cybernetic" or "kinetic" concepts; cf. Locker, 1964), it can be asked which point of view is the more general and fundamental one. In general terms, this is a question to be put to the Turing machine as a general automaton.

One consideration to the point (not, so far as we have seen, treated in automata theory) is the problem of "immense" numbers. The fundamental statement of automata theory is that happenings that can be defined in a finite number of "words" can be realized by an automaton (e.g., a formal neural network after McCulloch and Pitts, or a Turing machine) (von Neumann, 1951). The question lies in the term "finite." The automaton can, by definition, realize a finite series of events (however large), but not an infinite one. However, what if the number of steps required is "immense," i.e., not infinite, but for example transcending the number of particles in the universe (estimated to be of the order 10^{80}) or of events possible in the time span of the universe or some of its subunits (according to Elsasser's, 1966, proposal, a number whose logarithm is a large number)? Such immense numbers appear in many system problems with exponentials, factorials and other explosively increasing functions. They are encountered in systems even of a moderate number of components with strong (nonnegligible) interactions (cf. Ashby, 1964). To "map" them in a Turing machine, a tape of "immense" length would be required, i.e., one exceeding not only practical but physical limitations.

Consider, for a simple example, a directed graph of N points (Rapoport, 1959b). Between each pair an arrow may exist or may not exist (two possibilities). There are therefore $2^{N(N-1)}$ different ways to connect N points. If N is only 5, there are over a million ways to connect the points. With $N = 20$, the number of ways exceeds the estimated number of atoms in the universe. Similar problems arise, e.g., with possible connections between neurons (estimated of the order of 10 billion in the human brain) and with the genetic code (Repge, 1962). In the code, there is a minimum of 20 "words" (nucleotide triplets) spelling the twenty amino acids (actually 64); the code may contain some millions of units. This gives $20^{1,000,000}$ possibilities. Supposing the Laplacean spirit is to find out the functional value of every combination; he would have to make such number of probes, but there are only 10^{80} atoms in the universe. Let us presume (Repge, 1962) that 10^{30} cells are present on the earth at a certain point of time. Further assuming a new cell generation every minute would give, for an age of the earth of 15 billion years (10^{16} minutes), 10^{46} cells in total. To be sure to obtain a maximum number, 10^{20} life-bearing planets may be assumed. Then, in the whole universe, there certainly would be no more than 10^{66} living beings—which is a great number but far from being "immense." The estimate can be made with different assumptions (e.g., number of possible proteins or enzymes) but with essentially the same result.

Again, according to Hart (1959), human invention can be conceived as new combinations of previously existing elements. If so, the opportunity for new inventions will increase roughly as a function of the number of possible permutations and combinations of available elements, which means that its increase will be a factorial of the number of elements. Then the rate of acceleration of social change is itself accelerating so that in many cases not a logarithmic but a log-log acceleration will be found in cultural change. Hart presents interesting curves showing that increases in human speed, in killing areas of weapons, in life expectation, etc., actually followed such expression, i.e., the rate of cultural growth is not exponential or compound interest, but is super-acceleration in the way of a log-log curve. In a general way, limits of automata will appear if regulation in a system is directed not against one or a limited number of disturbances, but against "arbitrary" disturbances, i.e., an indefinite number of situations that could not possibly have been "foreseen"; this is widely the case in embryonic (e.g., experiments of Driesch) and neural (e.g., experiments of Lashley) regulations. Regulation here results from interaction of many components (cf. discussion in

Jeffries, 1951, pp. 32ff.). This, as von Neumann himself conceded, seems connected with the "self-restoring" tendencies of organismic as contrasted to technological systems; expressed in more modern terms, with their open-system nature which is not provided even in the abstract model of automaton such as a Turing machine.

It appears therefore that, as vitalists like Driesch have emphasized long ago, the mechanistic conception, even taken in the modern and generalized form of a Turing automaton, founders with regulations after "arbitrary" disturbances, and similarly in happenings where the number of steps required is "immense" in the sense indicated. Problems of realizability appear even apart from the paradoxes connected with infinite sets.

The above considerations pertain particularly to a concept or complex of concepts which indubitably is fundamental in the general theory of systems: that of *hierarchic order*. We presently "see" the universe as a tremendous hierarchy, from elementary particles to atomic nuclei, to atoms, molecules, high-molecular compounds, to the wealth of structures (electron and light-microscopic) between molecules and cells (Weiss, 1962b), to cells, organisms and beyond to supra-individual organizations. One attractive scheme of hierarchic order (there are others) is that of Boulding (Table 1.2). A similar hierarchy is found both in "structures" and in "functions." In the last resort, structure (i.e., order of parts) and function (order of processes) may be the very same thing: in the physical world matter dissolves into a play of energies, and in the biological world structures are the expression of a flow of processes. At present, the system of physical laws relates mainly to the realm between atoms and molecules (and their summation in macrophysics), which obviously is a slice of a much broader spectrum. Laws of organization and organizational forces are insufficiently known in the subatomic and the supermolecular realms. There are inroads into both the subatomic world (high energy physics) and the supermolecular (physics of high molecular compounds); but these are apparently at the beginnings. This is shown, on the one hand, by the present confusion of elementary particles, on the other, by the present lack of physical understanding of structures seen under the electronmicroscope and the lack of a "grammar" of the genetic code (cf. p. 153).

A general theory of hierarchic order obviously will be a mainstay of general systems theory. Principles of hierarchic order can be stated in verbal language (Koestler, 1967; in press); there are semimathematical ideas (Simon, 1965) connected with matrix theory, and formulations in terms of mathematical logic (Woodger, 1930–31). In graph theory hierarchic order is expressed by the "tree," and relational aspects of hierarchies can be represented in this way. But the problem is much broader and deeper: The question of hierarchic order is intimately connected with those of differentiation, evolution, and the measure of organization which does not seem to be expressed adequately in terms either of energetics (negative entropy) or of information theory (bits) (cf. pp. 150ff.). In the last resort, as mentioned, hierarchic order and dynamics may be the very same, as Koestler has nicely expressed in his simile of "The Tree and the Candle."

Thus there is an array of system models, more or less progressed and elaborate. Certain concepts, models and principles of general systems theory, such as hierarchic order, progressive differentiation, feedback, systems characteristics defined by set and graph theory, etc., are applicable broadly to material, psychological and sociocultural systems; others, such as open system defined by the exchange of matter, are limited to certain subclasses. As practice in applied systems analysis shows, diverse system models will have to be applied according to the nature of the case and operational criteria.

Table 1.2

An Informal Survey of Main Levels in the Hierarchy of Systems.
Partly in pursuance of Boulding, 1956b

LEVEL	DESCRIPTION AND EXAMPLES	THEORY AND MODELS
Static structures	Atoms, molecules, crystals, biological structures from the electron-microscopic to the macroscopic level	E.g. structural formulas of chemistry; crystallography; anatomical descriptions
Clock works	Clocks, conventional machines in general, solar systems	Conventional physics such as laws of mechanics (Newtonian and Einsteinian) and others
Control mechanisms	Thermostat, servo-mechanisms, homeostatic mechanism in organisms	Cybernetics; feedback and information theory

LEVEL	DESCRIPTION AND EXAMPLES	THEORY AND MODELS
Open systems	Flame, cells and organisms in general	(a) Expansion of physical theory to systems maintaining themselves in flow of matter (metabolism). (b) Information storage in genetic code (DNA). Connection of (a) and (b) presently unclear
Lower organisms	"Plant-like" organisms: Increasing differentiation of system (so-called "division of labor" in the organism); distinction of reproduction and functional individual ("germ track and soma")	Theory and models almost lacking
Animals	Increasing importance of traffic in information (evolution of receptors, nervous systems); learning; beginnings of consciousness	Beginnings in automata theory (S-R relations), feedback (regulatory phenomena), autonomous behavior (relaxation oscillations), etc.
Man	Symbolism; past and future, self and world, self-awareness, etc., as consequences; communication by language, etc.	Incipient theory of symbolism
Socio-cultural systems	Populations of organisms (humans included); symbol-determined communities (cultures) in man only	Statistical and dynamic laws in population dynamics, sociology, economics, possibly history. Beginnings of a theory of cultural systems.
Symbolic systems	Language, logic, mathematics, sciences, arts, morals, etc.	Algorithms of symbols (e.g. mathematics, grammar); "rules of the game" such as in visual arts, music, etc.

NB.—This survey is impressionistic and intuitive with no claim for logical rigor. Higher levels as a rule presuppose lower ones (e.g. life phenomena those at the physico-chemical level, socio-cultural phenomena the level of human activity, etc.); but the relation of levels requires clarification in each case (cf. problems such as open system and genetic code as apparent prerequisites of "life"; relation of "conceptual" to "real" systems, etc.). In this sense, the survey suggests both the limits of reductionism and the gaps in actual knowledge.

2 The Meaning of General System Theory

The Quest for a General System Theory

Modern science is characterized by its ever-increasing specialization, necessitated by the enormous amount of data, the complexity of techniques and of theoretical structures within every field. Thus science is split into innumerable disciplines continually generating new subdisciplines. In consequence, the physicist, the biologist, the psychologist and the social scientist are, so to speak, encapsulated in their private universes, and it is difficult to get word from one cocoon to the other.

This, however, is opposed by another remarkable aspect. Surveying the evolution of modern science, we encounter a surprising phenomenon. Independently of each other, similar problems and conceptions have evolved in widely different fields.

It was the aim of classical physics eventually to resolve natural phenomena into a play of elementary units governed by "blind" laws of nature. This was expressed in the ideal of the Laplacean spirit which, from the position and momentum of particles, can predict the state of the universe at any point in time. This mechanistic view was not altered but rather reinforced when deterministic laws in physics were replaced by statistical laws. According to Boltzmann's derivation of the second principle of thermodynamics, physical events are directed toward states of maximum probability, and physical laws, therefore, are essentially "laws of disorder," the outcome of unordered, statistical events. In contrast to this mechanistic view, however, problems of wholeness, dynamic interaction and organization have appeared in the various branches of modern physics. In the Heisenberg relation and quantum physics, it became impossible to resolve phenomena into local events; problems of order and organization appear whether the question is the structure of atoms, the architecture of proteins, or interaction phenomena in thermodynamics. Similarly biology, in the mechanistic conception, saw its goal in the resolution of life phenomena into atomic entities and partial processes. The living organism was resolved into cells, its activities into physiological and ultimately physicochemical processes, behavior into unconditioned and conditioned reflexes, the substratum of heredity into particulate genes, and so forth. In contradistinction, the organismic conception is basic for modern biology. It is necessary to study not only parts and processes in isolation, but also to solve the decisive problems found in the organization and order unifying them, resulting from dynamic interaction of parts, and making the behavior of parts different when studied in isolation or within the whole. Again, similar trends appeared in psychology. While classical association psychology attempted to resolve mental phenomena into elementary units—psychological atoms as it were—such as elementary sensations and the like, gestalt psychology showed the existence and primacy of psychological wholes which are not a summation of elementary units and are governed by dynamic laws. Finally, in the social sciences the concept of society as a sum of individuals as social atoms, e.g., the model of Economic Man, was replaced by the tendency to consider society, economy, nation as a whole superordinated to its parts. This implies the great problems of planned economy, of the deification of nation and state, but also reflects new ways of thinking.

This parallelism of general cognitive principles in different fields is even more impressive when one considers the fact that those developments took place in mutual independence and mostly without any knowledge of work and research in other fields.

There is another important aspect of modern science. Up to recent times, exact science, the corpus of laws of nature, was almost identical with theoretical physics. Few attempts to state exact laws in nonphysical fields have gained recognition. However, the impact of and progress in the biological, behavioral and social sciences seem to make necessary an expansion of our conceptual schemes in order to allow for systems of laws in fields where application of physics is not sufficient or possible.

Such a trend towards generalized theories is taking place in many fields and in a variety of ways. For example, an elaborate theory of the dynamics of biological populations, the struggle for existence and biological equilibria, has developed, starting with the pioneering work by Lotka and Volterra. The theory operates with biological notions, such as individuals, species, coefficients of competition, and the like. A similar procedure is applied in quantitative economics and econometrics. The models and families of equations applied in the latter happen to be similar to those of Lotka or, for that matter, of chemical kinetics, but the model of interacting entities and forces is again at a different level. To take another example: living organisms are essentially open systems, i.e., systems exchanging matter with their environment. Conventional physics and physical chemistry dealt with closed systems, and only in recent years has theory been expanded to include irreversible processes, open systems, and states of disequilibrium. If, however, we want to apply the model of open systems to, say, the phenomena of animal growth, we automatically come to a generalization of theory referring not to physical but to biological units. In other words, we are dealing with generalized systems. The same is true of the fields of cybernetics and information theory which have gained so much interest in the past few years.

Thus, there exist models, principles, and laws that apply to generalized systems or their subclasses, irrespective of their particular kind, the nature of their component elements, and the relations or "forces" between them. It seems legitimate to ask for a theory, not of systems of a more or less special kind, but of universal principles applying to systems in general.

In this way we postulate a new discipline called *General System Theory*. Its subject matter is the formulation and derivation of those principles which are valid for "systems" in general.

The meaning of this discipline can be circumscribed as follows. Physics is concerned with systems of different levels of generality.

It extends from rather special systems, such as those applied by the engineer in the construction of a bridge or of a machine; to special laws of physical disciplines, such as mechanics or optics; to laws of great generality, such as the principles of thermodynamics that apply to systems of intrinsically different nature, mechanic, caloric, chemical or whatever. Nothing prescribes that we have to end with the systems traditionally treated in physics. Rather, we can ask for principles applying to systems in general, irrespective of whether they are of physical, biological or sociological nature. If we pose this question and conveniently define the concept of system, we find that models, principles, and laws exist which apply to generalized systems irrespective of their particular kind, elements, and the "forces" involved.

A consequence of the existence of general system properties is the appearance of structural similarities or isomorphisms in different fields. There are correspondences in the principles that govern the behavior of entities that are, intrinsically, widely different. To take a simple example, an exponential law of growth applies to certain bacterial cells, to populations of bacteria, of animals or humans, and to the progress of scientific research measured by the number of publications in genetics or science in general. The entities in question, such as bacteria, animals, men, books, etc., are completely different, and so are the causal mechanisms involved. Nevertheless, the mathematical law is the same. Or there are systems of equations describing the competition of animal and plant species in nature. But it appears that the same systems of equations apply in certain fields in physical chemistry and in economics as well. This correspondence is due to the fact that the entities concerned can be considered, in certain respects, as "systems," i.e., complexes of elements standing in interaction. The fact that the fields mentioned, and others as well, are concerned with "systems," leads to a correspondence in general principles and even in special laws when the conditions correspond in the phenomena under consideration.

In fact, similar concepts, models and laws have often appeared in widely different fields, independently and based upon totally different facts. There are many instances where identical principles were discovered several times because the workers in one field were unaware that the theoretical structure required was already well developed in some other field. General system theory will go a long way towards avoiding such unnecessary duplication of labor.

System isomorphisms also appear in problems which are recalcitrant to quantitative analysis but are nevertheless of great intrinsic interest. There are, for example, isomorphies between biological systems and "epiorganisms" (Gerard) like animal communities and human societies. Which principles are common to the several levels of organization and so may legitimately be transferred from one level to another, and which are specific so that transfer leads to dangerous fallacies? Can societies and civilizations be considered as systems?

It seems, therefore, that a general theory of systems would be a useful tool providing, on the one hand, models that can be used in, and transferred to, different fields, and safeguarding, on the other hand, from vague analogies which often have marred the progress in these fields.

There is, however, another and even more important aspect of general system theory. It can be paraphrased by a felicitous formulation due to the well-known mathematician and founder of information theory, Warren Weaver. Classical physics, Weaver said, was highly successful in developing the theory of unorganized complexity. Thus, for example, the behavior of a gas is the result of the unorganized and individually untraceable movements of innumerable molecules; as a whole it is governed by the laws of thermodynamics. The theory of unorganized complexity is ultimately rooted in the laws of chance and probability and in the second law of thermodynamics. In contrast, the fundamental problem today is that of organized complexity. Concepts like those of organization, wholeness, directiveness, teleology, and differentiation are alien to conventional physics. However, they pop up everywhere in the biological, behavioral and social sciences, and are, in fact, indispensable for dealing with living organisms or social groups. Thus a basic problem posed to modern science is a general theory of organization. General system theory is, in principle, capable of giving exact definitions for such concepts and, in suitable cases, of putting them to quantitative analysis.

If we have briefly indicated what general system theory means, it will avoid misunderstanding also to state what it is not. It has been objected that system theory amounts to no more than the trivial fact that mathematics of some sort can be applied to different sorts of problems. For example, the law of exponential growth is applicable to very different phenomena, from radioactive decay to the extinction of human populations with insufficient reproduction. This, however, is so because the formula is one of the simplest differential equations, and can therefore be applied to quite different things. Therefore, if so-called isomorphic laws of growth occur in entirely different processes, it has no more significance than the fact that elementary arithmethic is applicable to all countable objects, that 2 plus 2 make 4, irrespective of whether the counted objects are apples, atoms or galaxies.

The answer to this is as follows. Not just in the example quoted by way of simple illustration, but in the development of system theory, the question is not the application of well-known mathematical expressions. Rather, problems are posed that are novel and partly far from solution. As mentioned, the method of classical science was most appropriate for phenomena that either can be resolved into isolated causal chains, or are the statistical outcome of an "infinite" number of chance processes, as is true of statistical mechanics, the second principle of thermodynamics and all laws deriving from it. The classical modes of thinking, however, fail in the case of interaction of a large but limited number of elements or processes. Here those problems arise which are circumscribed by such notions as wholeness, organization and the like, and which demand new ways of mathematical thinking.

Another objection emphasizes the danger that general system theory may end up in meaningless analogies. This danger indeed exists. For example, it is a widespread idea to look at the state or the nation as an organism on a superordinate level. Such a theory, however, would constitute the foundation for a totalitarian state, within which the human individual appears like an insignificant cell in an organism or an unimportant worker in a beehive.

But general system theory is not a search for vague and superficial analogies. Analogies as such are of little value since besides similarities between phenomena, dissimilarities can always be found as well. The isomorphism under discussion is more than mere analogy. It is a consequence of the fact that, in certain respects, corresponding abstractions and conceptual models can be applied to different phenomena. Only in view of these aspects will system laws apply. This is not different from the general procedure in science. It is the same situation as when the law of gravitation applies to Newton's apple, the planetary system, and tidal phenomena. This means that in view of certain limited aspects a theoretical system, that of mechanics, holds true; it does not mean that there is a particular resemblance between apples, planets, and oceans in a great number of other aspects.

A third objection claims that system theory lacks explanatory value. For example, certain aspects of organic purposiveness, such as the so-called equifinality of developmental processes (p. 40), are open to system-theoretical interpretation. Nobody, however, is today capable of defining in detail the processes leading from an animal ovum to an organism with its myriad of cells, organs, and highly complicated functions.

Here we should consider that there are degrees in scientific explanation, and that in complex and theoretically little-developed fields we have to be satisfied with what the economist Hayek has justly termed "explanation in principle." An example may show what is meant.

Theoretical economics is a highly developed system, presenting elaborate models for the processes in question. However, professors of economics, as a rule, are not millionaires. In other words, they can explain economic phenomena well "in principle" but they are not able to predict fluctuations in the stock market with respect to certain shares or dates. Explanation in principle, however, is better than none at all. If and when we are able to insert the necessary parameters, system-theoretical explanation "in principle" becomes a theory, similar in structure to those of physics.

Aims of General System Theory

We may summarize these considerations as follows.

Similar general conceptions and viewpoints have evolved in various disciplines of modern science. While in the past, science tried to explain observable phenomena by reducing them to an interplay of elementary units investigatable independently of each other, conceptions appear in contemporary science that are concerned with what is somewhat vaguely termed "wholeness," i.e., problems of organization, phenomena not resolvable into local events, dynamic interactions manifest in the difference of behavior of parts when isolated or in a higher configuration, etc.; in short, "systems" of various orders not understandable by investigation of their respective parts in isolation. Conceptions and problems of this nature have appeared in all branches of science, irrespective of whether inanimate things, living organisms, or social phenomena are the object of study. This correspondence is the more striking because the developments in the individual sciences were mutually independent, largely unaware of each other, and based upon different facts and contradicting philosophies. They indicate a general change in scientific attitude and conceptions.

Not only are general aspects and viewpoints alike in different sciences; frequently we find formally identical or isomorphic laws in different fields. In many cases, isomorphic laws hold for certain classes or subclasses of "systems," irrespective of the nature of the entities involved. There appear to exist general system laws which apply to any system of a certain type, irrespective of the particular properties of the system and of the elements involved.

These considerations lead to the postulate of a new scientific discipline which we call general system theory. Its subject matter is formulation of principles that are valid for "systems" in general, whatever the nature of their component elements and the relations or "forces" between them.

General system theory, therefore, is a general science of "wholeness" which up till now was considered a vague, hazy, and semimetaphysical concept. In elaborate form it would be a logicomathematical discipline, in itself purely formal but applicable to the various empirical sciences. For sciences concerned with "organized wholes," it would be of similar significance to that which probability theory has for sciences concerned with "chance events"; the latter, too, is a formal mathematical discipline which can be applied to most diverse fields, such as thermodynamics, biological and medical experimentation, genetics, life insurance statistics, etc.

This indicates major aims of general system theory:

(1) There is a general tendency towards integration in the various sciences, natural and social.

(2) Such integration seems to be centered in a general theory of systems.

(3) Such theory may be an important means for aiming at exact theory in the nonphysical fields of science.

(4) Developing unifying principles running "vertically" through the universe of the individual sciences, this theory brings us nearer to the goal of the unity of science.

(5) This can lead to a much-needed integration in scientific education.

A remark as to the delimitation of the theory here discussed seems to be appropriate. The term and program of a general system theory was introduced by the present author a number of years ago. It has turned out, however, that quite a large number of workers in various fields had been led to similar conclusions and ways of approach. It is suggested, therefore, to maintain this name which is now coming into general use, be it only as a convenient label.

It looks, at first, as if the definition of systems as "sets of elements standing in interrelation" is so general and vague that not much can be learned from it. This, however, is not true. For example, systems can be defined by certain families of differential equations and if, in the usual way of mathematical reasoning, more specified conditions are introduced, many important properties can be found of systems in general and more special cases (cf. Chapter 3).

The mathematical approach followed in general system theory is not the only possible or most general one. There are a number of related modern approaches, such as information theory, cybernetics, game, decision, and net theories, stochastic models, operations research, to mention only the most important ones. However, the fact that differential equations cover extensive fields in the physical, biological, economical, and probably also the behavioral sciences, makes them a suitable access to the study of generalized systems.

I am now going to illustrate general system theory by way of some examples.

Closed and Open Systems: Limitations of Conventional Physics

My first example is that of closed and open systems. Conventional physics deals only with closed systems, i.e., systems which are considered to be isolated from their environment. Thus, physical chemistry tells us about the reactions, their rates, and the chemical equilibria eventually established in a closed vessel where a number of reactants is brought together. Thermodynamics expressly declares that its laws apply only to closed systems. In particular, the second principle of thermodynamics states that, in a closed system, a certain quantity, called entropy, must increase to a maximum, and eventually the process comes to a stop at a state of equilibrium. The second principle can be formulated in different ways, one being that entropy is a measure of probability, and so a closed system tends to a state of most probable distribution. The most probable distribution, however, of a mixture, say, of red and blue glass beads, or of molecules having different velocities, is a state of complete disorder; having separated all red beads on one hand, and all blue ones on the other, or having, in a closed space, all fast molecules, that is, a high temperature on the right side, and all slow ones, a low temperature, at the left, is a highly improbable state of affairs. So the tendency towards maximum entropy or the most probable distribution is the tendency to maximum disorder.

However, we find systems which by their very nature and definition are not closed systems. Every living organism is essentially an open system. It maintains itself in a continuous inflow and outflow, a building up and breaking down of components, never being, so long as it is alive, in a state of chemical and thermodynamic equilibrium but maintained in a so-called steady state which is distinct from the latter. This is the very essence of that fundamental phenomenon of life which is called metabolism, the chemical processes within living cells. What now? Obviously, the conventional formulations of physics are, in principle, inapplicable to the living organism *qua* open system and steady state, and we may well suspect that many characteristics of living systems which are paradoxical in view of the laws of physics are a consequence of this fact.

It is only in recent years that an expansion of physics, in order

to include open systems, has taken place. This theory has shed
light on many obscure phenomena in physics and biology, and
has also led to important general conclusions of which I will
mention only two.

The first is the principle of equifinality. In any closed system,
the final state is unequivocally determined by the initial conditions: e.g., the motion in a planetary system where the positions
of the planets at a time t are unequivocally determined by their
positions at a time t_0. Or in a chemical equilibrium, the final
concentrations of the reactants naturally depend on the initial
concentrations. If either the initial conditions or the process is
altered, the final state will also be changed. This is not so in
open systems. Here, the same final state may be reached from
different initial conditions and in different ways. This is what is
called equifinality, and it has a significant meaning for the
phenomena of biological regulation. Those who are familiar with
the history of biology will remember that it was just equifinality
that led the German biologist Driesch to embrace vitalism, i.e.,
the doctrine that vital phenomena are inexplicable in terms of
natural science. Driesch's argument was based on experiments on
embryos in early development. The same final result, a normal
individual of the sea urchin, can develop from a complete ovum,
from each half of a divided ovum, or from the fusion product of
two whole ova. The same applies to embryos of many other
species, including man, where identical twins are the product of
the splitting of one ovum. Equifinality, according to Driesch,
contradicts the laws of physics, and can be accomplished only by
a soul-like vitalistic factor which governs the processes in foresight of the goal, the normal organism to be established. It can
be shown, however, that open systems, insofar as they attain a
steady state, must show equifinality, so the supposed violation
of physical laws disappears (cf. pp. 132f.).

Another apparent contrast between inanimate and animate
nature is what sometimes was called the violent contradiction between Lord Kelvin's degradation and Darwin's evolution, between
the law of dissipation in physics and the law of evolution in
biology. According to the second principle of thermodynamics,
the general trend of events in physical nature is toward states
of maximum disorder and levelling down of differences, with the
so-called heat death of the universe as the final outlook, when

all energy is degraded into evenly distributed heat of low temperature, and the world process comes to a stop. In contrast, the living world shows, in embryonic development and in evolution, a transition towards higher order, heterogeneity, and organization. But on the basis of the theory of open systems, the apparent contradiction between entropy and evolution disappears. In all irreversible processes, entropy must increase. Therefore, the change of entropy in closed systems is always positive; order is continually destroyed. In open systems, however, we have not only production of entropy due to irreversible processes, but also import of entropy which may well be negative. This is the case in the living organism which imports complex molecules high in free energy. Thus, living systems, maintaining themselves in a steady state, can avoid the increase of entropy, and may even develop towards states of increased order and organization.

From these examples, you may guess the bearing of the theory of open systems. Among other things, it shows that many supposed violations of physical laws in living nature do not exist, or rather that they disappear with the generalization of physical theory. In a generalized version the concept of open systems can be applied to nonphysical levels. Examples are its use in ecology and the evolution towards a climax formation (Whittacker), in psychology where "neurological systems" were considered as "open dynamic systems" (Krech), in philosophy where the trend toward "trans-actional" as opposed to "self-actional" and "inter-actional" viewpoints closely corresponds to the open system model (Bentley).

Information and Entropy

Another development which is closely connected with system theory is that of the modern theory of communication. It has often been said that energy is the currency of physics, just as economic values can be expressed in dollars or pounds. There are, however, certain fields of physics and technology where this currency is not readily acceptable. This is the case in the field of communication which, due to the development of telephones, radio, radar, calculating machines, servomechanisms and other devices, has led to the rise of a new branch of physics.

The general notion in communication theory is that of information. In many cases, the flow of information corresponds to a flow of energy, e.g., if light waves emitted by some objects reach the eye or a photoelectric cell, elicit some reaction of the organism or some machinery, and thus convey information. However, examples can easily be given where the flow of information is opposite to the flow of energy, or where information is transmitted without a flow of energy or matter. The first is the case in a telegraph cable, where a direct current is flowing in one direction, but information, a message, can be sent in either direction by interrupting the current at one point and recording the interruption at another. For the second case, think of the photoelectric door openers as they are installed in many supermarkets: the shadow, the cutting off of light energy, informs the photocell that somebody is entering, and the door opens. So information, in general, cannot be expressed in terms of energy.

There is, however, another way to measure information, namely, in terms of decisions. Take the game of Twenty Questions, where we are supposed to find out an object by receiving simple "yes" or "no" answers to our questions. The amount of information conveyed in one answer is a decision between two alternatives, such as animal or nonanimal. With two questions, it is possible to decide for one out of four possibilities, e.g., mammal—nonmammal, or flowering plant—nonflowering plant. With three answers, it is a decision out of eight, etc. Thus, the logarithm at the base 2 of the possible decisions can be used as a measure of information, the unit being the so-called binary unit or bit. The information contained in two answers is $\log_2 4 = 2$ bits, of three answers, $\log_2 8 = 3$ bits, etc. This measure of information happens to be similar to that of entropy or rather negative entropy, since entropy also is defined as a logarithm of probability. But entropy, as we have already heard, is a measure of disorder; hence negative entropy or information is a measure of order or of organization since the latter, compared to distribution at random, is an improbable state.

A second central concept of the theory of communication and control is that of feedback. A simple scheme for feedback is the following (Fig. 2.1). The system comprises, first, a receptor or "sense organ," be it a photoelectric cell, a radar screen, a thermometer, or a sense organ in the biological meaning. The message may be, in technological devices, a weak current, or, in a

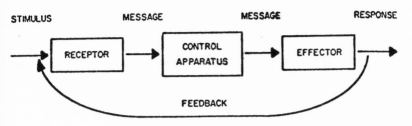

STIMULUS   MESSAGE   MESSAGE   RESPONSE

RECEPTOR → CONTROL APPARATUS → EFFECTOR

FEEDBACK

Fig. 2.1. Simple feedback scheme.

living organism, represented by nerve conduction, etc. Then there is a center recombining the incoming messages and transmitting them to an effector, consisting of a machine like an electromotor, a heating coil or solenoid, or of a muscle which responds to the incoming message in such a way that there is power output of high energy. Finally, the functioning of the effector is monitored back to the receptor, and this makes the system self-regulating, i.e., guarantees stabilization or direction of action.

Feedback arrangements are widely used in modern technology for the stabilization of a certain action, as in thermostats or in radio receivers; or for the direction of actions towards a goal where the aberration from that goal is fed back, as information, till the goal or target is reached. This is the case in self-propelled missiles which seek their target, anti-aircraft fire control systems, ship-steering systems, and other so-called servomechanisms.

There is indeed a large number of biological phenomena which correspond to the feedback model. First, there is the phenomenon of so-called homeostasis, or maintenance of balance in the living organism, the prototype of which is thermoregulation in warm-blooded animals. Cooling of the blood stimulates certain centers in the brain which "turn on" heat-producing mechanisms of the body, and the body temperature is monitored back to the center so that temperature is maintained at a constant level. Similar homeostatic mechanisms exist in the body for maintaining the constancy of a great number of physicochemical variables. Furthermore, feedback systems comparable to the servomechanisms of technology exist in the animal and human body for the regulation of actions. If we want to pick up a pencil, a report is made to the central nervous system of the distance by which we have failed to grasp the pencil in the first instance; this information is then fed back to the central nervous system so that the motion is controlled till the aim is reached.

So a great variety of systems in technology and in living nature follow the feedback scheme, and it is well-known that a new discipline, called Cybernetics, was introduced by Norbert Wiener to deal with these phenomena. The theory tries to show that mechanisms of a feedback nature are the base of teleological or purposeful behavior in man-made machines as well as in living organisms, and in social systems.

It should be borne in mind, however, that the feedback scheme is of a rather special nature. It presupposes structural arrangements of the type mentioned. There are, however, many regulations in the living organism which are of essentially different nature, namely, those where the order is effectuated by a dynamic interplay of processes. Recall, e.g., embryonic regulations where the whole is reestablished from the parts in equifinal processes. It can be shown that the *primary* regulations in organic systems, i.e., those which are most fundamental and primitive in embryonic development as well as in evolution, are of the nature of dynamic interaction. They are based upon the fact that the living organism is an open system, maintaining itself in, or approaching a steady state. Superposed are those regulations which we may call *secondary*, and which are controlled by fixed arrangements, especially of the feedback type. This state of affairs is a consequence of a general principle of organization which may be called progressive mechanization. At first, systems—biological, neurological, psychological or social—are governed by dynamic interaction of their components; later on, fixed arrangements and conditions of constraint are established which render the system and its parts more efficient, but also gradually diminish and eventually abolish its equipotentiality. Thus, dynamics is the broader aspect, since we can always arrive from general system laws to machinelike function by introducing suitable conditions of constraint, but the opposite is not possible.

Causality and Teleology

Another point I would like to mention is the change the

scientific world picture has undergone in the past few decades. In the world view called mechanistic, which was born of classical physics of the nineteenth century, the aimless play of the atoms, governed by the inexorable laws of causality, produced all phenomena in the world, inanimate, living, and mental. No room was left for any directiveness, order, or telos. The world of the organisms appeared a product of chance, accumulated by the senseless play of random mutations and selection; the mental world as a curious and rather inconsequential epiphenomenon of material events.

The only goal of science appeared to be analytical, i.e., the splitting up of reality into ever smaller units and the isolation of individual causal trains. Thus, physical reality was split up into mass points or atoms, the living organism into cells, behavior into reflexes, perception into punctual sensations, etc. Correspondingly, causality was essentially one-way: one sun attracts one planet in Newtonian mechanics, one gene in the fertilized ovum produces such and such inherited character, one sort of bacterium produces this or that disease, mental elements are lined up, like the beads in a string of pearls, by the law of association. Remember Kant's famous table of the categories which attempts to systematize the fundamental notions of classical science: it is symptomatic that the notions of interaction and of organization were only space-fillers or did not appear at all.

We may state as characteristic of modern science that this scheme of isolable units acting in one-way causality has proved to be insufficient. Hence the appearance, in all fields of science, of notions like wholeness, holistic, organismic, gestalt, etc., which all signify that, in the last resort, we must think in terms of systems of elements in mutual interaction.

Similarly, notions of teleology and directiveness appeared to be outside the scope of science and to be the playground of mysterious, supernatural or anthropomorphic agencies; or else, a pseudoproblem, intrinsically alien to science, and merely a misplaced projection of the observer's mind into a nature governed by purposeless laws. Nevertheless, these aspects exist, and you cannot conceive of a living organism, not to speak of behavior and human society, without taking into account what variously and rather loosely is called adaptiveness, purposiveness, goal-seeking and the like.

It is characteristic of the present view that these aspects are taken seriously as a legitimate problem for science; moreover, we can well indicate models simulating such behavior.

Two such models we have already mentioned. One is equifinality, the tendency towards a characteristic final state from different initial states and in different ways, based upon dynamic interaction in an open system attaining a steady state; the second, feedback, the homeostatic maintenance of a characteristic state or the seeking of a goal, based upon circular causal chains and mechanisms monitoring back information on deviations from the state to be maintained or the goal to be reached. A third model for adaptive behavior, a "design for a brain," was developed by Ashby, who incidentally started with the same mathematical definitions and equations for a general system as were used by the present author. Both writers have developed their systems independently and, following different lines of interest, have arrived at different theorems and conclusions. Ashby's model for adaptiveness is, roughly, that of step functions defining a system, i.e., functions which, after a certain critical value is passed, jump into a new family of differential equations. This means that, having passed a critical state, the system starts off in a new way of behavior. Thus, by means of step functions, the system shows adaptive behavior by what the biologist would call trial and error: it tries different ways and means, and eventually settles down in a field where it no longer comes into conflict with critical values of the environment. Such a system adapting itself by trial and error was actually constructed by Ashby as an electromagnetic machine, called the homeostat.

I am not going to discuss the merits and shortcomings of these models of teleological or directed behavior. What should be stressed, however, is the fact that teleological behavior directed towards a characteristic final state or goal is not something off limits for natural science and an anthropomorphic misconception of processes which, in themselves, are undirected and accidental. Rather it is a form of behavior which can well be defined in scientific terms and for which the necessary conditions and possible mechanisms can be indicated.

What Is Organization?

Similar considerations apply to the concept of organization.

Organization also was alien to the mechanistic world. The problem did not appear in classical physics, mechanics, electrodynamics, etc. Even more, the second principle of thermodynamics indicated destruction of order as the general direction of events. It is true that this is different in modern physics. An atom, a crystal, or a molecule are organizations, as Whitehead never failed to emphasize. In biology, organisms are, by definition, organized things. But although we have an enormous amount of data on biological organization, from biochemistry to cytology to histology and anatomy, we do not have a theory of biological organization, i.e., a conceptual model which permits explanation of the empirical facts.

Characteristic of organization, whether of a living organism or a society, are notions like those of wholeness, growth, differentiation, hierarchical order, dominance, control, competition, etc. Such notions do not appear in conventional physics. System theory is well capable of dealing with these matters. It is possible to define such notions within the mathematical model of a system; moreover, in some respects, detailed theories can be developed which deduce, from general assumptions, the special cases. A good example is the theory of biological equilibria, cyclic fluctuations, etc., as initiated by Lotka, Volterra, Gause and others. It will certainly be found that Volterra's biological theory and the theory of quantitative economics are isomorphic in many respects.

There are, however, many aspects of organizations which do not easily lend themselves to quantitative interpretation. This difficulty is not unknown in natural science. Thus, the theory of biological equilibria or that of natural selection are highly developed fields of mathematical biology, and nobody doubts that they are legitimate, essentially correct, and an important part of the theory of evolution and of ecology. It is hard, however, to apply them in the field because the parameters chosen, such as selective value, rate of destruction and generation and the like, cannot easily be measured. So we have to content ourselves with an "explanation in principle," a qualitative argument which, however, may lead to interesting consequences.

As an example of the application of general system theory to human society, we may quote a recent book by Boulding, entitled *The Organizational Revolution*. Boulding starts with a general model of organization and states what he calls Iron Laws which hold good for any organization. Such Iron Laws are, for example, the Malthusian law that the increase of a population is, in general, greater than that of its resources. Then there is a law of optimum size of organizations: the larger an organization grows, the longer is the way of communication and this, depending on the nature of the organization, acts as a limiting factor and does not allow an organization to grow beyond a certain critical size. According to the law of instability, many organizations are not in a stable equilibrium but show cyclic fluctuations which result from the interaction of subsystems. This, incidentally, could probably be treated in terms of the Volterra theory, Volterra's so-called first law being that of periodic cycles in populations of two species, one of which feeds at the expense of the other. The important law of oligopoly states that, if there are competing organizations, the instability of their relations and hence the danger of friction and conflicts increases with the decrease of the number of those organizations. Thus, so long as they are relatively small and numerous, they muddle through in some way of coexistence. But if only a few or a competing pair are left, as is the case with the colossal political blocks of the present day, conflicts become devastating to the point of mutual destruction. The number of such general theorems for organization can easily be enlarged. They are well capable of being developed in a mathematical way, as was actually done for certain aspects.

General System Theory and the Unity of Science

Let me close these remarks with a few words about the general implications of interdisciplinary theory.

The integrative function of general system theory can perhaps be summarized as follows. So far, the unification of science has been seen in the reduction of all sciences to physics, the final resolution of all phenomena into physical events. From our point of view, unity of science gains a more realistic aspect. A unitary conception of the world may be based, not upon the possibly futile and certainly farfetched hope finally to reduce all levels of reality to the level of physics, but rather on the isomorphy of laws in different fields. Speaking in what has been called the "formal" mode, i.e., looking at the conceptual constructs of science, this means structural uniformities of the schemes we are applying. Speaking in "material" language, it means that the world, i.e., the total of observable events, shows structural uniformities, manifesting themselves by isomorphic traces of order in the different levels or realms.

We come, then, to a conception which in contrast to reductionism, we may call perspectivism. We cannot reduce the biological, behavioral, and social levels to the lowest level, that of the constructs and laws of physics. We can, however, find constructs and possibly laws within the individual levels. The world is, as Aldous Huxley once put it, like a Neapolitan ice cream cake where the levels—the physical, the biological, the social and the moral universe—represent the chocolate, strawberry, and vanilla layers. We cannot reduce strawberry to chocolate—the most we can say is that possibly in the last resort, all is vanilla, all mind or spirit. The unifying principle is that we find organization at all levels. The mechanistic world view, taking the play of physical particles as ultimate reality, found its expression in a civilization which glorifies physical technology that has led eventually to the catastrophes of our time. Possibly the model of the world as a great organization can help to reinforce the sense of reverence for the living which we have almost lost in the last sanguinary decades of human history.

General System Theory in Education:
The Production of Scientific Generalists

After this sketchy outline of the meaning and aims of general system theory, let me try to answer the question of what it may contribute to integrative education. In order not to appear partisan, I give a few quotations from authors who were not themselves engaged in the development of general system theory.

A few years ago, a paper, entitled "The Education of Scientific Generalists," was published by a group of scientists including the engineer Bode, the sociologist Mosteller, the mathematician Tukey, and the biologist Winsor. The authors emphasized the "need for a simpler, more unified approach to scientific problems." They wrote:

> We often hear that "one man can no longer cover a broad enough field" and that "there is too much narrow specializa-

tion." ... We need a simpler, more unified approach to scientific problems, we need men who practice science—not a particular science, in a word, we need scientific generalists (Bode *et al.*, 1949).

The authors then make clear how and why generalists are needed in fields such as physical chemistry, biophysics, the application of chemistry, physics, and mathematics to medicine, and they continue:

Any research group needs a generalist, whether it is an institutional group in a university or a foundation, or an industrial group.... In an engineering group, the generalist would naturally be concerned with system problems. These problems arise whenever parts are made into a balanced whole (Bode *et al.*, 1949).

In a symposium of the Foundation for Integrated Education, Professor Mather (1951) discussed "Integrative Studies for General Education." He stated:

One of the criticisms of general education is based upon the fact that it may easily degenerate into the mere presentation of information picked up in as many fields of enquiry as there is time to survey during a semester or a year.... If you were to overhear several senior students talking, you might hear one of them say "our professors have stuffed us full, but what does it all mean?" ... More important is the search for basic concepts and underlying principles that may be valid throughout the entire body of knowledge.

In answer to what these basic concepts may be, Mather states:

Very similar general concepts have been independently developed by investigators who have been working in widely different fields. These correspondences are all the more significant because they are based upon totally different facts. The men who developed them were largely unaware of each other's work. They started with conflicting philosophies and yet have reached remarkably similar conclusions....

Thus conceived, [Mather concludes], integrative studies would prove to be an essential part of the quest for an understanding of reality.

No comments seem to be necessary. Conventional education in physics, biology, psychology or the social sciences treats them as separate domains, the general trend being that increasingly smaller subdomains become separate sciences, and this process is repeated to the point where each specialty becomes a triflingly small field, unconnected with the rest. In contrast, the educational demands of training "Scientific Generalists" and of developing interdisciplinary "basic principles" are precisely those general system theory tries to fill. They are not a mere program or a pious wish since, as we have tried to show, such theoretical structure is already in the process of development. In this sense, general system theory seems to be an important headway towards interdisciplinary synthesis and integrated education.

Science and Society

However, if we speak of education, we do not mean solely scientific values, i.e., communication and integration of facts. We also mean ethical values, contributing to the development of personality. Is there something to be gained from the viewpoints we have discussed? This leads to the fundamental problem of the value of science in general and the behavioral and social sciences in particular.

An often-used argument about the value of science and its impact upon society and the welfare of mankind runs something like this. Our knowledge of the laws of physics is excellent, and consequently our technological control of inanimate nature almost unlimited. Our knowledge of biological laws is not so far advanced, but sufficient to allow for a good amount of biological technology in modern medicine and applied biology. It has extended the life expectancy far beyond the limits allotted to human beings in earlier centuries or even decades. The application of the modern methods of scientific agriculture, husbandry, etc., would well suffice to sustain a human population far surpassing the present one of our planet. What is lacking, however, is knowledge of the laws of human society, and consequently a sociological technology. So the achievements of physics are put to use for ever more efficient destruction; we have famines in vast parts of the world while harvests rot or are destroyed in other parts; war and indiscriminate annihilation of human life,

culture, and means of sustenance are the only way out of un-
controlled fertility and consequent overpopulation. They are the
outcome of the fact that we know and control physical forces
only too well, biological forces tolerably well, and social forces
not at all. If, therefore, we would have a well-developed science
of human society and a corresponding technology, it would be
the way out of the chaos and impending destruction of our
present world.

This seems to be plausible enough and is, in fact, but a modern
version of Plato's precept that only if the rulers are philosophers,
humanity will be saved. There is, however, a catch in the argument. We have a fair idea what a scientifically controlled world
would look like. In the best case, it would be like Aldous
Huxley's *Brave New World*, in the worst, like Orwell's *1984*.
It is an empirical fact that scientific achievements are put just
as much, or even more, to destructive as constructive use. The
sciences of human behavior and society are no exception. In
fact, it is perhaps the greatest danger of the systems of modern
totalitarianism that they are so alarmingly up-to-date not only
in physical and biological, but also in psychological technology.
The methods of mass suggestion, of the release of the instincts of
the human beast, of conditioning and thought control are developed to highest efficacy; just because modern totalitarianism
is so terrifically scientific, it makes the absolutism of former
periods appear a dilettantish and comparatively harmless make-
shift. Scientific control of society is no highway to Utopia.

The Ultimate Precept: Man as the Individual

We may, however, conceive of a scientific understanding of
human society and its laws in a somewhat different and more
modest way. Such knowledge can teach us not only what human
behavior and society have in common with other organizations,
but also what is their uniqueness. Here the main tenet will be:
Man is not only a political animal; he is, before and above all,
an individual. The real values of humanity are not those which
it shares with biological entities, the function of an organism or
a community of animals, but those which stem from the individual mind. Human society is not a community of ants or
termites, governed by inherited instinct and controlled by the

laws of the superordinate whole; it is based upon the achievements of the individual and is doomed if the individual is made a cog in the social machine. This, I believe, is the ultimate precept a theory of organization can give: not a manual for dictators of any denomination more efficiently to subjugate human beings by the scientific application of Iron Laws, but a warning that the Leviathan of organization must not swallow the individual without sealing its own inevitable doom.

3 Some System Concepts in Elementary Mathematical Consideration

The System Concept

In dealing with complexes of "elements," three different kinds of distinction may be made—i.e., 1. according to their *number*; 2. according to their *species*; 3. according to the *relations* of elements. The following simple graphical illustration may clarify this point (FIG. **3.1**) with *a* and *b* symbolizing various complexes.

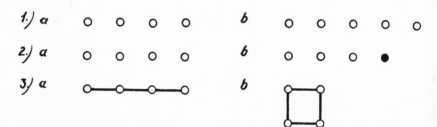

Fig. *3.1.* See text.

In cases 1 and 2, the complex may be understood as the (cf. pp. 66ff.) sum of elements considered in isolation. In case 3, not only the elements should be known, but also the relations between them. Characteristics of the first kind may be called *summative*, of the second kind *constitutive*. We can also say that summative characteristics of an element are those which are the same within and outside the complex; they may therefore be obtained by means of summation of characteristics and behavior of elements as known in isolation. Constitutive characteristics are those which are dependent on the specific relations within the complex; for understanding such characteristics we therefore must know not only the parts, but also the relations.

Physical characteristics of the first type are, for example, weight or molecular weight (sum of weights or atomic weights respectively), heat (considered as sum of movements of the molecules), etc. An example of the second kind are chemical characteristics (e.g., isomerism, different characteristics of compounds with the same gross composition but different arrangement of radicals in the molecule).

The meaning of the somewhat mystical expression, "the whole is more than the sum of parts" is simply that constitutive characteristics are not explainable from the characteristics of isolated parts. The characteristics of the complex, therefore, compared to those of the elements, appear as "new" or "emergent." If, however, we know the total of parts contained in a system and the relations between them, the behavior of the system may be derived from the behavior of the parts. We can also say: While we can conceive of a sum as being composed gradually, a system as total of parts with its interrelations has to be conceived of as being composed instantly.

Physically, these statements are trivial; they could become problematic and lead to confused conceptions in biology, psychology and sociology only because of a misinterpretation of the mechanistic conception, the tendency being towards resolution of phenomena into independent elements and causal chains, while interrelations were bypassed.

In rigorous development, general system theory would be of an axiomatic nature; that is, from the notion of "system" and a suitable set of axioms propositions expressing system properties and principles would be deduced. The following considerations are much more modest. They merely illustrate some system principles by formulations which are simple and intuitively accessible, without attempt at mathematical rigor and generality.

A system can be defined as a set of elements standing in interrelations. Interrelation means that elements, p, stand in relations, R, so that the behavior of an element p in R is different from its behavior in another relation, R'. If the behaviors in R and R' are not different, there is no interaction, and the elements behave independently with respect to the relations R and R'.

A system can be defined mathematically in various ways. For illustration, we choose a system of simultaneous differential equations. Denoting some measure of elements, p_i $(i = 1, 2, \ldots n)$, by Q_i, these, for a finite number of elements and in the simplest case, will be of the form:

$$
\left.
\begin{aligned}
\frac{dQ_1}{dt} &= f_1 \left(Q_1, Q_2, \ldots Q_n \right) \\
\frac{dQ_2}{dt} &= f_2 \left(Q_1, Q_2, \ldots Q_n \right) \\
&\cdots\cdots\cdots\cdots\cdots\cdots\cdots \\
\frac{dQ_n}{dt} &= f_n \left(Q_1, Q_2, \ldots Q_n \right)
\end{aligned}
\right\}
\tag{3.1}
$$

Change of any measure Q_i therefore is a function of all Q's, from Q_1 to Q_n; conversely, change of any Q_i entails change of all other measures and of the system as a whole.

Systems of equations of this kind are found in many fields and represent a general principle of kinetics. For example, in *Simultankinetik* as developed by Skrabal (1944, 1949), this is the general expression of the law of mass action. The same system was used by Lotka (1925) in a broad sense, especially with respect to demographic problems. The equations for biocoenotic systems, as developed by Volterra, Lotka, D'Ancona, Gause and others, are special cases of equation (3.1). So are the equations used by Spiegelman (1945) for kinetics of cellular processes and the theory of competition within an organism. G. Werner (1947) has stated a similar though somewhat more general system (considering the system as continuous, and using therefore partial differential equations with respect to x, y, z, and t) as the basic law of pharmacodynamics from which the various laws of drug action can be derived by introducing the relevant special conditions.

Such a definition of "system" is, of course, by no means general. It abstracts from spatial and temporal conditions, which would be expressed by partial differential equations. It also abstracts from a possible dependence of happenings on the previous history of the system ("hysteresis" in a broad sense); consideration of this would make the system into integro-differential equations as discussed by Volterra (1931; cf. also d'Ancona, 1939) and Donnan (1937). Introduction of such equations would have a definite meaning: The system under consideration would be not only a spatial but also a temporal whole.

Notwithstanding these restrictions, equation (3.1) can be used for discussing several general system properties. Although nothing is said about the nature of the measures Q_i or the functions f_i—i.e., about the relations or interactions within the system— certain general principles can be deduced.

There is a condition of stationary state, characterized by dis-appearance of the changes dQ_i/dt

$$f_1 = f_2 = \ldots f_n = 0 \qquad (3.2)$$

By equating to zero we obtain n equations for n variables, and by solving them obtain the values:

$$\left. \begin{array}{l} Q_1 = Q_1{}^* \\ Q_2 = Q_2{}^* \\ \cdots \cdots \cdots \\ Q_a = Q_n{}^* \end{array} \right\} \qquad (3.3)$$

These values are constants, since in the system, as presupposed, the changes disappear. In general, there will be a number of stationary states, some stable, some instable.

We may introduce new variables:

$$Q_i = Q_i{}^* - Q_i{}' \qquad (3.4)$$

and reformulate system (3.1):

$$\left. \begin{array}{l} \dfrac{dQ_1{}'}{dt} = f_1{}' \left(Q_1{}', Q_2{}', \ldots Q_n{}' \right) \\[2mm] \dfrac{dQ_2{}'}{dt} = f_2{}' \left(Q_1{}', Q_2{}', \ldots Q_n{}' \right) \\[2mm] \cdots \cdots \cdots \cdots \cdots \cdots \cdots \cdots \\[2mm] \dfrac{dQ_n{}'}{dt} = f_n{}' \left(Q_1{}', Q_2{}', \ldots Q_n{}' \right) \end{array} \right\} \qquad (3.5)$$

Let us assume that the system can be developed in Taylor series:

$$\frac{dQ_1'}{dt} = a_{11} Q_1' + a_{12} Q_2' + \ldots$$

$$a_{1n} Q_n' + a_{111} Q_1'^2 + a_{112} Q_1' Q_2' + a_{122} Q_2'^2 + \ldots$$

$$\frac{dQ_2'}{dt} = a_{21} Q_1' + a_{22} Q_2' + \ldots$$

$$a_{2n} Q_n' + a_{211} Q_1'^2 + a_{212} Q_1' Q_2' + a_{222} Q_2'^2 + \ldots \qquad (3.6)$$

$$\cdots \cdots \cdots \cdots \cdots \cdots \cdots \cdots \cdots \cdots \cdots \cdots \cdots \cdots$$

$$\frac{dQ_n'}{dt} = a_{n1} Q_1' + a_{n2} Q_2' + \ldots$$

$$a_{nn} Q_n' + a_{n11} Q_1'^2 + a_{n12} Q_1' Q_2' + a_{n22} Q_2'^2 + \ldots$$

A general solution of this system of equations is:

$$Q_1' = G_{11} e^{\lambda_1 t} + G_{12} e^{\lambda_2 t} + \ldots G_{1n} e^{\lambda_n t} + G_{111} e^{2\lambda_1 t} + \ldots$$

$$Q_2' = G_{21} e^{\lambda_1 t} + G_{22} e^{\lambda_2 t} + \ldots G_{2n} e^{\lambda_n t} + G_{211} e^{2\lambda_1 t} + \ldots$$

$$\cdots \cdots \cdots \cdots \cdots \cdots \cdots \cdots \cdots \cdots \cdots \cdots \cdots \qquad (3.7)$$

$$Q_n' = G_{n1} e^{\lambda_1 t} + G_{n2} e^{\lambda_2 t} + \ldots G_{nn} e^{\lambda_n t} + G_{n11} e^{2\lambda_1 t} + \ldots$$

where the G are constants and the λ the roots of the characteristic equation:

$$\begin{vmatrix} a_{11} - \lambda & a_{12} & a_{1n} \\ a_{21} & a_{22} - \lambda & a_{2n} \\ \cdots \cdots \cdots \cdots \cdots \cdots \cdots \cdots \\ a_{n1} & a_{n2} & a_{nn} - \lambda \end{vmatrix} = 0 \qquad (3.8)$$

The roots λ may be real or imaginary. By inspection of equations (3.7) we find that if all λ are real and negative (or, if complex, negative in their real parts), Q_i', with increasing time, approach 0 because $e^{-\infty} = 0$; since, however, according to (3.5) $Q_i = Q_i^* - Q_i'$, the Q_i thereby obtain the stationary values Q_i^*. In this case the equilibrium is *stable*, since in a sufficient period of time the system comes as close to the stationary state as possible.

However, if one of the λ is positive or 0, the equilibrium is *unstable*, i.e., the system will move away from equilibrium.

If finally some λ are positive and complex, the system contains periodic terms since the exponential function for complex exponents takes the form:

$$e^{(a-ib)t} = e^{at} \,(\cos bt - i \sin bt).$$

In this case there will be *periodic fluctuations*, which generally are damped.

For illustration, consider the simplest case, $n = 2$, a system consisting of two kinds of elements:

$$\left. \begin{aligned} \frac{dQ_1}{dt} &= f_1 \, (Q_1, \, Q_2) \\ \frac{dQ_2}{dt} &= f_2 \, (Q_1, \, Q_2) \end{aligned} \right\} \tag{3.9}$$

Again provided that the functions can be developed into Taylor series, the solution is:

$$\left. \begin{aligned} Q_1 &= Q_1{}^* - G_{11}e^{\lambda_1 t} - G_{12}e^{\lambda_2 t} - G_{111}e^{2\lambda_1 t} - \cdots \\ Q_2 &= Q_2{}^* - G_{21}e^{\lambda_1 t} - G_{22}e^{\lambda_2 t} - G_{211}e^{2\lambda_1 t} - \cdots \end{aligned} \right\} \tag{3.10}$$

with $Q_1{}^*$, $Q_2{}^*$ as stationary values of Q_1, Q_2, obtained by setting $f_1 = f_2 = 0$; the G's integration constants; and the λ's roots of the characteristic equation:

$$\begin{vmatrix} a_{11} - \lambda & a_{12} \\ a_{21} & a_{22} - \lambda \end{vmatrix} = 0,$$

or developed:

$$(a_{11} - \lambda) \, (a_{22} - \lambda) - a_{12}a_{21} = 0,$$
$$\lambda^2 - \lambda C + D = 0,$$
$$\lambda = \frac{C}{2} \pm \sqrt{- D + \left(\frac{C}{2}\right)^2},$$

with

$$C = a_{11} + a_{22}; \; D = a_{11}a_{22} - a_{12}a_{21}.$$

In the case:

$$C < 0, D > 0, E = C^2 - 4D > 0,$$

both solutions of the characteristic equation are negative. Therefore a *node* is given; the system will approach a stable stationary state $(Q_1{}^*, \, Q_2{}^*)$ as $e^{-\infty} = 0$, and therefore the second and following terms continually decrease (FIG. **3.2**).

In the case:

$$C < 0, D > 0, E = C^2 - 4D < 0,$$

both solutions of the characteristic equation are complex with negative real part. In this case we have a *loop*, and point $(Q_1, \, Q_2)$ tends towards $(Q_1{}^*, \, Q_2{}^*)$ describing a spiral curve.

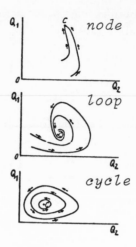

Fig. 3.2. See text.

In the case:

$$C = 0, \ D > 0, \ E < 0,$$

both solutions are imaginary, therefore the solution contains periodic terms; there will be oscillations or cycles around the stationary values. Point (Q_1, Q_2) describes a *closed curve* around $(Q_1{}^*, Q_2{}^*)$.

In the case:

$$C > 0, D < 0, E > 0,$$

both solutions are positive, and there is no stationary state.

Growth

Equations of this type are found in a variety of fields, and we can use system (3.1) to illustrate the formal identity of system laws in various realms, in other words, to demonstrate the existence of a general system theory.

This may be shown for the simplest case—i.e., the system consisting of elements of only one kind. Then the system of equations is reduced to the single equation:

$$\frac{dQ}{dt} = f(Q), \tag{3.11}$$

which may be developed into a Taylor series:

$$\frac{dQ}{dt} = a_1 Q + a_{11} Q^2 + \ldots \tag{3.12}$$

This series does not contain an absolute term in the case in which there is no "spontaneous generation" of elements. Then dQ/dt must disappear for $Q = 0$, which is possible only if the absolute term is equal to zero.

The simplest possibility is realized when we retain only the first term of the series:

$$\frac{dQ}{dt} = a_1 Q, \tag{3.13}$$

This signifies that the growth of the system is directly proportional to the number of elements present. Depending on whether the constant a_1 is positive or negative, the growth of the system is positive or negative, and the system increases or decreases. The solution is:

$$Q = Q_0 e^{a_1 t}, \tag{3.14}$$

Q_0 signifying the number of elements at $t = 0$. This is the *exponential law* (FIG. **3.3**) found in many fields.

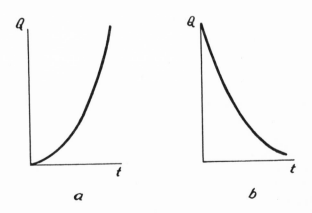

Fig. 3.3. Exponential curves.

In mathematics, the exponential law is called the "law of natural growth," and with $(a_1 > 0)$ is valid for the growth of capital by compound interest. Biologically, it applies to the individual growth of certain bacteria and animals. Sociologically, it is valid for the unrestricted growth of plant or animal populations, in the simplest case for the increase of bacteria when each individual divides into two, these into four, etc. In social science, it is called the law of Malthus and signifies the unlimited growth of a population, whose birth rate is higher than its death rate. It also describes the growth of human knowledge as measured by the number of textbook pages devoted to scientific discoveries, or the number of publications on drosophila (Hersh, 1942). With negative constant $(a_1 < 0)$, the exponential law applies to radioactive decay, to the decomposition of a chemical compound in monomolecular reaction, to the killing of bacteria by rays or poison, the loss of body substance by hunger in a multicellular organism, the rate of extinction of a population in which the death rate is higher than the birth rate, etc.

Going back to equation (3.12) and retaining two terms, we have:

$$\frac{dQ}{dt} = a_1 Q + a_{11} Q^2 \tag{3.15}$$

A solution of this equation is:

$$Q = \frac{a_1 C e^{a_1 t}}{1 - a_{11} C e^{a_1 t}} \tag{3.16}$$

Keeping the second term has an important consequence. The simple exponential (3.14) shows an infinite increase; taking into account the second term, we obtain a curve which is sigmoid and attains a limiting value. This curve is the so-called *logistic curve* (FIG. 3.4), and is also of wide application.

In chemistry, this is the curve of an autocatalytical reaction, i.e., a reaction, in which the reaction product obtained accelerates its own production. In sociology, it is the law of Verhulst (1838) describing the growth of human populations with limited resources.

Mathematically trivial as these examples are, they illustrate a point of interest for the present consideration, namely the fact that certain laws of nature can be arrived at not only on

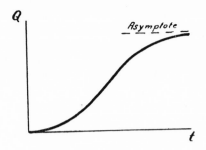

Fig. 3.4. Logistic curve.

the basis of experience, but also in a purely formal way. The equations discussed signify no more than that the rather general system of equation (3.1), its development into a Taylor series and suitable conditions have been applied. In this sense such laws are "*a priori*," independent from their physical, chemical, biological, sociological, etc., interpretation. In other words, this shows the existence of a general system theory which deals with formal characteristics of systems, concrete facts appearing as their special applications by defining variables and parameters. In still other terms, such examples show a formal uniformity of nature.

Competition

Our system of equations may also indicate competition between parts.

The simplest possible case is, again, that all coefficients $(a_{j \neq i}) = 0$, — i.e., that the increase in each element depends only on this element itself. Then we have, for two elements:

$$\left. \begin{aligned} \frac{dQ_1}{dt} &= a_1 Q_1 \\[2mm] \frac{dQ_2}{dt} &= a_2 Q_2 \end{aligned} \right\} \tag{3.17}$$

or

$$\left. \begin{aligned} Q_1 &= c_1 e^{a_1 t} \\[2mm] Q_2 &= c_2 e^{a_2 t} \end{aligned} \right\} \tag{3.18}$$

Eliminating time, we obtain:

$$t = \frac{\ln Q_1 - \ln c_1}{a_1} = \frac{\ln Q_2 - \ln c_2}{a_2}, \qquad (3.19)$$

and

$$Q_1 = bQ_2{}^\alpha \qquad (3.20)$$

with
$$\alpha = a_1/a_2, \quad b = c_1/c_2{}^\alpha.$$

This is the equation known in biology as the *allometric equation*. In this discussion, the simplest form of growth of the parts —viz., the exponential—has been assumed (**3.17** and **3.18**). The allometric relation holds, however, also for somewhat more complicated cases, such as growth according to the parabola, the logistic, the Gompertz function, either strictly or as an approximation (Lumer, 1937).

The allometric equation applies to a wide range of morphological, biochemical, physiological and phylogenetic data. It means that a certain characteristic, Q_1, can be expressed as a power function of another characteristic, Q_2. Take, for instance, morphogenesis. Then the length or weight of a certain organ, Q_1, is, in general, an allometric function of the size of another organ, or of the total length or weight of the organism in question, Q_2. The meaning of this becomes clear if we write equations (**3.17**) in a slightly different form:

$$\frac{dQ_1}{dt} \cdot \frac{1}{Q_1} : \frac{dQ_2}{dt} \cdot \frac{1}{Q_2} = \alpha, \qquad (3.21)$$

or

$$\frac{dQ_1}{dt} = \alpha \cdot \frac{Q_1}{Q_2} \cdot \frac{dQ_2}{dt} \qquad (3.22)$$

Equation (**3.21**) states that the relative growth rates (i.e., increase calculated as a percentage of actual size) of the parts under consideration, Q_1 and Q_2, stand in a constant proportion throughout life, or during a life cycle for which the allometric equation holds. This rather astonishing relation (because of the immense complexity of growth processes it would seem, at first, unlikely that the growth of parts is governed by an algebraic equation of such simplicity) is explained by equation (3.22). According to this equation, it can be interpreted as a result of a process of distribution. Take Q_2 for the whole organism; then equation (3.22) states that the organ Q_1 takes, from the increase resulting from the metabolism of the total organism $\left(\dfrac{dQ_2}{dt}\right)$, a share which is proportional to its actual proportion to the latter $\left(\dfrac{Q_1}{Q_2}\right)$. α is a partition coefficient indicating the capacity of the organ to seize its share. If $a_1 > a_2$—i.e., if the growth intensity of Q_1 is greater than that of Q_2—then $\alpha = \dfrac{a_1}{a_2} > 1$; the organ captures more than other parts; it grows therefore more rapidly than these or with positive allometry. Conversely, if $a_1 < a_2$—i.e., $a < 1$—the organ grows more slowly, or shows negative allometry. Similarly, the allometric equation applies to biochemical changes in the organism, and to physiological functions. For instance, basal metabolism increases, in wide groups of animals, with $a = 2/3$ with respect to body weight if growing animals of the same species, or animals of related species, are compared; this means that basal metabolism is, in general, a surface function of body weight. In certain cases, such as insect larvae and snails, $\alpha = 1$, i.e. basal metabolism is proportional to weight itself.

In sociology, the expression in question is *Pareto's law* (1897) of the distribution of income within a nation, whereby $Q_1 = bQ_2{}^\alpha$, with Q_1 = number of individuals gaining a certain income, Q_2 = amount of the income, and b and α constants. The explanation is similar to that given above, substituting for "increase of the total organism" the national income, and for "distribution constant" the economic abilities of the individuals concerned.

The situation becomes more complex if interactions between the parts of the system are assumed—i.e., if $a_{j \neq i} \neq 0$. Then we come to systems of equations such as those studied by Volterra (1931) for competition among species, and, correspondingly, by Spiegelman (1945) for competition within an organism. Since these cases are fully discussed in the literature we shall not enter into a detailed discussion. Only one or two points of general interest may be mentioned.

It is an interesting consequence that, in Volterra's equations, competition of two species for the same resources is, in a way, more fatal than a predator-prey relation—i.e., partial annihilation of one species by the other. Competition eventually leads to the extermination of the species with the smaller growth capacity; a predator-prey relation only leads to periodic oscillation of the numbers of the species concerned around a mean value. These relations have been stated for biocoenotic systems, but it may well be that they have also sociological implications.

Another point of philosophical interest should be mentioned. If we are speaking of "systems," we mean "wholes" or "unities." Then it seems paradoxical that, with respect to a whole, the concept of competition between its parts is introduced. In fact, however, these apparently contradictory statements both belong to the essentials of systems. Every whole is based upon the competition of its elements, and presupposes the "struggle between parts" (Roux). The latter is a general principle of organization in simple physico-chemical systems as well as in organisms and social units, and it is, in the last resort, an expression of the *coincidentia oppositorum* that reality presents.

Wholeness, Sum, Mechanization, Centralization

The concepts just indicated have often been considered to describe characteristics only of living beings, or even to be a proof of vitalism. In actual fact they are formal properties of systems.

(1) Let us assume again that the equations (3.1) can be developed into Taylor series:

$$\frac{dQ_1}{dt} = a_{11}Q_1 + a_{12}Q_2 + \ldots a_{1n}Q_n + a_{111}Q^2_1 + \ldots \quad \text{(3.23)}$$

We see that any change in some quantity, Q_1, is a function of the quantities of all elements, Q_1 to Q_n. On the other hand, a change in a certain Q_i causes a change in all other elements and in the total system. The system therefore behaves as a *whole*, the changes in every element depending on all the others.

(2) Let the coefficients of the variables Q_j $(j \neq i)$ now become zero. The system of equations degenerates into:

$$\frac{dQ_i}{dt} = a_{i1}Q_i + a_{i11}Q_i^2 + \ldots \qquad (3.24)$$

This means that a change in each element depends only on that element itself. Each element can therefore be considered independent of the others. The variation of the total complex is the (physical) sum of the variations of its elements. We may call such behavior *physical summativity* or *independence*.

We may define summativity by saying that a complex can be built up, step by step, by putting together the first separate elements; conversely, the characteristics of the complex can be analyzed completely into those of the separate elements. This is true for those complexes which we may call "heaps," such as a heap of bricks or odds and ends, or for mechanical forces acting according to the parallelogram of forces. It does not apply to those systems which were called *Gestalten* in German. Take the most simple example: three electrical conductors have a certain charge which can be measured in each conductor separately. But if they are connected by wires, the charge in each conductor depends on the total constellation, and is different from its charge when insulated.

Though this is trivial from the viewpoint of physics, it is still necessary to emphasize the non-summative character of physical and biological systems because the methodological attitude has been, and is yet to a large extent, determined by the mechanistic program (von Bertalanffy, 1949a, 1960). In Lord Russell's book (1948), we find a rather astonishing rejection of the "concept of organism." This concept states, according to Russell, that the laws governing the behavior of the parts can be stated only by considering the place of the parts in the whole. Russell rejects this view. He uses the example of an eye, the function of which as a light receptor can be understood perfectly well if the eye is isolated and if only the internal physico-chemical reactions, and the incoming stimuli and outgoing nerve impulses, are taken into account. "Scientific progress has been made by analysis and artificial isolation. . . . It is therefore in any case prudent to adopt the mechanistic view as a working hypothesis, to be abandoned only where there is clear evidence against it. As regards biological phenomena, such evidence, so far, is entirely absent." It is true that the principles of summativity are applicable to the living organism to a certain extent. The beat of a heart, the twitch of a nerve-muscle preparation, the action potentials in a nerve are much the same if studied in isolation or within the organism as a whole. This applies to those phenomena we shall define later as occurring in highly "mechanized" partial systems. But Russell's statement is profoundly untrue with respect exactly to the basic and primary biological phenomena. If you take any realm of biological phenomena, whether embryonic development, metabolism, growth, activity of the nervous system, biocoenoses, etc., you will always find that the behavior of an element is different within the system from what it is in isolation. You cannot sum up the behavior of the whole from the isolated parts, and you have to take into account the relations between the various subordinated systems and the systems which are super-ordinated to them in order to understand the behavior of the parts. Analysis and artificial isolation are useful, but in no way sufficient, methods of biological experimentation and theory.

(3) *Summativity in the mathematical sense* means that the change in the total system obeys an equation of the same form as the equations for the parts. This is possible only when the functions on the right side of the equation contain linear terms only; a trivial case.

(4) There is a further case which appears to be unusual in physical systems but is common and basic in biological, psychological and sociological systems. This case is that in which the interactions between the elements decrease with time. In terms of our basic model equation (3.1), this means that the coefficients of the Q_i are not constant, but decrease with time. The simplest case will be:

$$\lim_{t \to \infty} a_{ij} = 0 \tag{3.25}$$

In this case the system passes from a state of wholeness to a state of independence of the elements. The primary state is that of a unitary system which splits up gradually into independent causal chains. We may call this *progressive segregation*.

As a rule, the organization of physical wholes, such as atoms, molecules, or crystals, results from the union of pre-existing elements. In contrast, the organization of biological wholes is built up by differentiation of an original whole which segregates into parts. An example is determination in embryonic development, when the germ passes from a state of equipotentiality to a state where it behaves like a mosaic or sum of regions which develop independently into definite organs. The same is true in the development and evolution of the nervous system and of behavior starting with actions of the whole body or of large regions and passing to the establishment of definite centers and localized reflex arcs, and for many other biological phenomena.

The reason for the predominance of segregation in living nature seems to be that segregation into subordinate partial systems implies an increase of complexity in the system. Such transition towards higher order presupposes a supply of energy, and energy is delivered continuously into the system only if the latter is an open system, taking energy from its environment. We shall come back to this question later on.

In the state of wholeness, a disturbance of the system leads to the introduction of a new state of equilibrium. If, however, the system is split up into individual causal chains, these go on independently. Increasing mechanization means increasing determination of elements to functions only dependent on themselves, and consequent loss of regulability which rests in the system as a whole, owing to the interrelations present. The smaller the interaction coefficients become, the more the respective terms Q_i can be neglected, and the more "machine-like" is the system—i.e., like a sum of independent parts.

This fact, which may be termed "progressive mechanization," plays an important role in biology. Primary, it appears, is behavior resulting from interaction within the system; secondarily, determination of the elements on actions dependent only on these elements, transition from behavior as a whole to summative behavior takes place. Examples are found in embryonic development, where originally the performance of each region depends on its position within the whole so that regulation following arbitrary disturbance is possible; later on, the embryonic regions are determined for one single performance—e.g., development of a certain organ. In the nervous system, similarly, certain parts become irreplaceable centers for certain—e.g., reflex—performances. Mechanization, however, is never complete in the biological realm; even though the organism is partly mechanized, it still remains a unitary system; this is the basis of regulation and of the interaction with changing demands of the environment. Similar considerations apply to social structures. In a primitive community every member can perform almost anything expected in his connection with the whole; in a highly differentiated community, each member is determined for a certain performance, or complex of performances. The extreme case is reached in certain insect communities, where the individuals are, so to speak, transformed into machines determined for certain performances. The determination of individuals into workers or soldiers in some ant communities by way of nutritional differences at certain stages amazingly resembles ontogenetic determination of germinal regions to a certain developmental fate.

In this contrast between wholeness and sum lies the tragical tension in any biological, psychological and sociological evolution. Progress is possible only by passing from a state of undifferentiated wholeness to differentiation of parts. This implies, however, that the parts become fixed with respect to a certain action. Therefore progressive segregation also means progressive mechanization. Progressive mechanization, however, implies loss of regulability. As long as a system is a unitary whole, a disturbance will be followed by the attainment of a new stationary state, due to the interactions within the system. The system is self-regulating. If, however, the system is split up into independent causal chains, regulability disappears. The partial processes will go on irrespective of each other. This is the behavior we find, for example, in embryonic development, determination going hand in hand with decrease of regulability.

Progress is possible only by subdivision of an initially unitary action into actions of specialized parts. This, however, means at the same time impoverishment, loss of performances still possible in the undetermined state. The more parts are specialized in a certain way, the more they are irreplaceable, and loss of parts may lead to the breakdown of the total system. To speak Aristotelian language, every evolution, by unfolding some potentiality, nips in the bud many other possibilities. We may find this in embryonic development as well as in phylogenetic specialization, or in specialization in science or daily life (von Bertalanffy, 1949a, 1960, pp. 42 ff.).

Behavior as a whole and summative behavior, unitary and elementalistic conceptions, are usually regarded as being antitheses. But it is frequently found that there is no opposition between them, but gradual transition from behavior as a whole to summative behavior.

(5) Connected with this is yet another principle. Suppose that the coefficients of one element, p_s, are large in all equations while the coefficients of the other elements are considerably smaller or even equal to zero. In this case the system may look like this:

$$
\left.
\begin{aligned}
\frac{dQ_1}{dt} &= a_{11}Q_1 + \ldots a_{1s}Q_s + \ldots \\
\frac{dQ_s}{dt} &= a_{s1}Q_s + \ldots \\
\frac{dQ_n}{dt} &= a_{ns}Q_s + \ldots a_{n1}Q_n + \ldots
\end{aligned}
\right\}
\qquad (3.26)
$$

if for simplicity we write the linear members only.

Then relationships are given which can be expressed in several ways. We may call the element p_s a *leading part,* or say that the system is *centered* around p_s. If the coefficients a_{is} of p_s in some or all equations are large while the coefficients in the equation of p_s itself are small, a small change in p_s will cause a considerable change in the total system. p_s may be then called a *trigger.* A small change in p_s will be "amplified" in the total system. From the energetic viewpoint, in this case we do not find "conservation causality" (*Erhaltungskausalität*) where the principle "*causa aequat effectum*" holds, but "instigation causality" (*Anstosskausalität*) (Mittasch, 1948), an energetically insignificant change in p_s causing a considerable change in the total system.

The principle of centralization is especially important in the biological realm. Progressive segregation is often connected with progressive centralization, the expression of which is the time-dependent evolution of a leading part—i.e., a combination of the schemes (3.25) and (3.26). At the same time, the principle of progressive centralization is that of progressive individualization. An "individual" can be defined as a centralized system. Strictly speaking this is, in the biological realm, a limiting case, only approached ontogenetically and phylogenetically, the organism growing through progressive centralization more and more unified and "more indivisible."

All these facts may be observed in a variety of systems. Nicolai Hartmann even demands centralization for every "dynamic structure." He therefore recognizes only a few kinds of structures, in the physical realm, those of smallest dimensions (the atom as a planetary system of electrons around a nucleus) and of large dimensions (planetary systems centralized by a sun). From the biological viewpoint, we would emphasize progressive mechanization and centralization. The primitive state is that where the behavior of the system results from the interactions of equipotential parts; progressively, subordination under dominant parts takes place. In embryology, for example, these are called organizers (Spemann); in the central nervous system, parts first are largely equipotential as in the diffuse nervous systems of lower animals; later on subordination to leading centers of the nervous system takes place.

Thus, similar to progressive mechanization a principle of progressive centralization is found in biology, symbolized by time-dependent formation of leading parts—i.e., a combination of schemes (3.25) and (3.26). This viewpoint casts light on an important, but not easily definable concept, that of the *individual*. "Individual" stands for "indivisible." Is it, however, possible to call a planarian or hydra an "individual" if these animals may be cut up into any number of pieces and still regenerate a complete animal? Double-headed hydras can easily be made by experiment; then the two heads may fight for a daphnia, although it is immaterial on which side the prey is caught; in any case it is swallowed to reach the common stomach where it is digested to the benefit of all parts. Even in higher organisms individuality is doubtful, at least in early development. Not only each half of a divided sea urchin embryo, but also the halves of a salamander embryo develop into complete animals; identical twins in man are, so to speak, the result of a Driesch experiment carried out by nature. Similar considerations apply to the behavior of animals: in lower animals tropotaxis may take place in the way of antagonistic action of the two halves of the body if they are appropriately exposed to stimuli; ascending the evolutionary scale, increasing centralization appears; behavior is not a resultant of partial mechanisms of equal rank but dominated and unified by the highest centers of the nervous system (cf. von Bertalanffy, 1937; pp. 131ff., 139ff.).

Thus strictly speaking, biological individuality does not exist, but only progressive individualization in evolution and development resulting from progressive centralization, certain parts gaining a dominant role and so determining behavior of the whole. Hence the principle of *progressive centralization* also constitutes *progressive individualization*. An individual is to be defined as a centered system, this actually being a limiting case approached in development and evolution so that the organism becomes more unified and "indivisible" (cf. von Bertalanffy, 1932; pp. 269ff.). In the psychological field, a similar phenomenon is the "centeredness" of gestalten, e.g., in perception; such centeredness appears necessary so that a psychic gestalt distinguishes itself from others. In contrast to the "principle of ranklessness" of association psychology, Metzger states (1941, p. 184) that "every psychic formation, object, process, experience down to the simplest gestalten of perception, exhibits a certain weight distribution and centralization; there is rank order, sometimes a derivative relationship, among its parts, loci, properties." The same applies again in the sociological realm: an amorphous mob has no "individuality"; in order that a social structure be distinguished from others, grouping around certain individuals is necessary. For this very reason, a biocoenosis like a lake or a forest is not an "organism," because an individual organism always is centered to a more or less large extent.

Neglect of the principle of progressive mechanization and centralization has frequently led to pseudoproblems, because only the limiting cases of independent and summative elements, or else complete interaction of equivalent elements were recognized, not the biologically important intermediates. This plays a role with respect to the problems of "gene" and "nervous center." Older genetics (not modern genetics any more) was inclined to consider the hereditary substance as a sum of corpuscular units determining individual characteristics or organs; the objection is obvious that a sum of macromolecules cannot produce the organized wholeness of the organism. The correct answer is that the genome as a whole produces the organism as a whole, certain genes, however, preeminently determining the direction of development of certain characters—i.e., acting as "leading parts." This is expressed in the insight that every hereditary trait is co-determined by many, perhaps all genes, and that every gene influences not one single trait but many, and possibly the total organism (polygeny of characteristics and polypheny of genes). In a similar way, in the function of the nervous system there was apparently the alternative of considering it either as a sum of mechanisms for the individual functions, or else as a homogeneous nervous net. Here, too, the correct conception is that any function ultimately results from interaction of all parts, but that certain parts of the central nervous system influence it decisively and therefore can be denoted as "centers" for that function.

(6) A more general (but less visualizable) formulation of what was said follows. If the change of Q_i be any function F_i of the Q_i and their derivatives in the space coordinates we have:

(2) If $\dfrac{\partial F_i}{\partial Q_j} = 0$, $i \neq j$: "independence";

(4) If $\dfrac{\partial F_i}{\partial Q_j} = f(t)$, $\lim\limits_{t \to \infty} \dfrac{\partial F_i}{\partial Q_j} = 0$: "progressive mechanization";

(5) If $\dfrac{\partial F_i}{\partial Q_s} \gg \dfrac{\partial F_i}{\partial Q_j}$, $j \neq s$, or even: $\dfrac{\partial F_i}{\partial Q_j} = 0$: Q_s is the "dominant part."

(7) The system concept as outlined asks for an important addition. Systems are frequently structured in a way so that their individual members again are systems of the next lower level. Hence each of the elements denoted by $Q_1, Q_2 \ldots Q_n$ is a system of elements $O_{i1}, O_{i2} \ldots O_{in}$, in which each system O is again definable by equations similar to those of (3.1):

$$\frac{dO_{ii}}{dt} = f_{ii} \left(O_{i1}, O_{i2}, \ldots O_{in} \right).$$

Such superposition of systems is called *hierarchical order*. For its individual levels, again the aspects of wholeness and summativity, progressive mechanization, centralization, finality, etc., apply.

Such hierarchical structure and combination into systems of ever higher order, is characteristic of reality as a whole and of fundamental importance especially in biology, psychology and sociology.

(8) An important distinction is that of *closed* and *open systems.* This will be discussed in Chapters 6–8.

Finality

As we have seen, systems of equations of the type considered may have three different kinds of solution. The system in question may asymptotically attain a stable stationary state with increasing time; it may never attain such state; or there may be periodic oscillations. In case the system approaches a stationary state, its variation can be expressed not only in terms of the actual conditions but also in terms of the distance from the stationary state. If $Q_i{}^*$ are the solutions for the stationary state, new variables:

$$Q_i = Q_i{}^* - Q_i{}'$$

can be introduced so that

$$\frac{dQ_1}{dt} = f(Q_1{}^* - Q_1{}') \, (Q_2{}^* - Q_2{}') \ldots (Q_n{}^* - Q_n{}') \qquad (3.27)$$

We may express this as follows. In case a system approaches a stationary state, changes occurring may be expressed not only in terms of actual conditions, but also in terms of the distance from the equilibrium state; the system seems to "aim" at an equilibrium to be reached only in the future. Or else, the happenings may be expressed as depending on a future final state.

It has been maintained for a long time that certain formulations in physics have an apparently finalistic character. This applies in two respects. Such teleology was especially seen in the *minimum principles of mechanics.* Already Maupertuis considered his minimum principle as proof that the world, where among many virtual movements the one leading to maximum effect with minimum effort is realized, is the "best of all worlds" and work of a purposeful creator. Euler made a similar remark: "Since the construction of the whole world is the most eminent and since it originated from the wisest creator, nothing is found in the world which would not show a maximum or minimum characteristic." A similar teleological aspect can be seen in Le Châtelier's principle in physical chemistry and in Lenz's rule of electricity. All these principles express that in case of disturbance, the system develops forces which counteract the disturbance and restore a state of equilibrium; they are derivations from the principle of minimum effect. Principles homologous to the principle of minimum action in mechanics can be construed for any type of system; thus Volterra (cf. d'Ancona, 1939; pp. 98ff.) has shown that a population dynamics homologous to mechanical dynamics can be developed where a similar principle of minimum action appears.

The conceptual error of an anthropomorphic interpretation is easily seen. The principle of minimum action and related principles simply result from the fact that, if a system reaches a state of equilibrium, the derivatives become zero; this implies that certain variables reach an extremum, minimum or maximum; only when these variables are denoted by anthropomorphic terms like effect, constraint, work, etc., an apparent teleology in physical processes emerges in physical action (cf. Bavink, 1944).

Finality can be spoken of also in the sense of *dependence on the future*. As can be seen from equation (3.27), happenings can, in fact, be considered and described as being determined not by actual conditions, but also by the final state to be reached. Secondly, this formulation is of a general nature; it does not only apply to mechanics, but to any kind of system. Thirdly, the question was frequently misinterpreted in biology and philosophy, so that clarification is fairly important.

Let us take, for a change, a growth equation formulated by the author (von Bertalanffy, 1934 and elsewhere). The equation is: $l = l^* - (l^* - l_0 e^{-kt})$ (cf. pp. 171ff.), where l represents the length of the animal at time t, l^* the final length, l_0 the initial length, and k a constant. This looks as if the length l of the animal at time t were determined by the final value l^* which will be reached only after infinitely long time. However, the final state (l^*) simply is an extremum condition obtained by equating the differential quotient to zero so that t disappears. In order to do so, we must first know the differential equation by which the process is actually determined. This differential equation is: $\frac{dl}{dt} = E - kl$ and states that growth is determined by a counteraction of processes of anabolism and catabolism, with parameters E and k respectively. In this equation, the process at time t is determined only by the actual conditions and no future state appears. By equating to zero, l^* is defined by E/k. The "teleological" final-value formula therefore is only a transformation of the differential equation indicating actual conditions. In other words, the directedness of the process towards a final state is not a process differing from causality, but another expression of it. The final state to be reached in future is not a *"vis a fronte"* mysteriously attracting the system, but only another expression for causal *"vires a tergo."* For this reason, physics makes ample use of such final-value formulas because the fact is mathematically clear and nobody attributes an anthropomorphic "foresight" of the goal to a physical system. Biologists, on the other hand, often regarded such formulas as somewhat uncanny, either fearing a hidden vitalism, or else considering such teleology or goal-directedness as "proof" for vitalism. For with respect to animate rather than to inanimate nature, we tend to compare finalistic processes with human foresight of the goal; while, in fact, we are dealing with obvious, even mathematically trivial relations.

This matter was frequently misinterpreted even by philosophers. From E. von Hartmann to modern authors like Kafka (1922) and myself, finality was defined as the reverse of causality, as dependence of the process on future instead of past conditions. This was frequently objected to because, according to this conception, a state A would depend on a state B in the future, an existent on a non-existent (e.g., Gross 1930; similarly Schlick). As the above shows, this formulation does not mean an inconceivable "action" of a not existent future, but merely a sometimes useful formulation of a fact which can be expressed in terms of causality.

Types of Finality

No detailed discussion of the problem of finality is intended here, but enumeration of several types may be useful. Thus we can distinguish:

(1) Static teleology or fitness, meaning that an arrangement seems to be useful for a certain "purpose." Thus a fur coat is fit to keep the body warm, and so are hairs, feathers, or layers of fat in animals. Thorns may protect plants against grazing cattle, or imitative colorations and mimicries

may be advantageous to protect animals against enemies.

(2) Dynamic teleology, meaning a directiveness of processes. Here different phenomena can be distinguished which are often confused:

(i)   Direction of events towards a final state which can be expressed as if the present behavior were dependent on that final state. Every system which attains a time-independent condition behaves in this way.

(ii) Directiveness based upon structure, meaning that an arrangement of structures leads the process in such way that a certain result is achieved. This is true, of course, of the function of man-made machines yielding products or performances as desired. In living nature we find a structural order of processes that in its complication widely surpasses all man-made machines. Such order is found from the function of macroscopic organs, such as the eye as a sort of camera or the heart as a pump, to microscopic cell structures responsible for metabolism, secretion, excitability, heredity and so forth. Whilst man-made machines work in such a way as to yield certain products and performances, for example, fabrication of airplanes or moving a railway train, the order of process in living systems is such as to maintain the system itself. An important part of these processes is represented by homeostasis (Canon)—i.e., those processes through which the material and energetical situation of the organism is maintained constant. Examples are the mechanisms of thermoregulation, of maintenance of osmotic pressure, of pH, of salt concentration, the regulation of posture and so forth. These regulations are governed, in a wide extent, by feedback mechanisms. Feedback means that from the output of a machine a certain amount is monitored back, as "information," to the input so as to regulate the latter and thus to stabilize or direct the action of the machine. Mechanisms of this kind are well known in technology, as, for instance, the governor of the steam-engine, self-steering missiles and other "servomechanisms." Feedback mechanisms appear to be responsible for a large part of organic regulations and phenomena of homeostasis, as recently emphasized by Cybernetics (Frank et al., 1948; Wiener, 1948).

(iii) There is, however, yet another basis for organic regulations. This is equifinality—i.e., the fact that the same final state can be reached from different initial conditions and in different ways. This is found to be the case in open systems, insofar as they attain a steady state. It appears that equifinality is responsible for the primary regulability of organic systems—i.e., for all those regulations which cannot be based upon predetermined structures or mechanisms, but on the contrary, exclude such mechanisms and were regarded therefore as arguments for vitalism.

(iv) Finally, there is true finality or purposiveness, meaning that the actual behavior is determined by the foresight of the goal. This is the original Aristotelian concept. It presupposes that the future goal is already present in thought, and directs the present action. True purposiveness is characteristic of human behavior, and it is connected with the evolution of the symbolism of language and concepts (von Bertalanffy, 1948a, 1965).

The confusion of these different types of finality is one of the factors responsible for the confusion occurring in epistemology and theoretical biology. In the field of man-made things, fitness (a) and teleological working of machines (b, ii) are, of course, due to a planning intelligence (b, iv). Fitness in organic structures (a) can presumably be explained by the causal play of random mutations and natural selection. This explanation is, however, much less plausible for the origin of the very complicated organic mechanisms and feedback systems (b, ii). Vitalism is essentially the attempt to explain organic directiveness (b, ii and iii) by means of intelligence in foresight of the goal (b, iv). This leads, methodologically, beyond the limits of natural science, and is empirically unjustified, since we have, even in the most astonishing phenomena of regulation or instinct, no justification for, but most definite reasons against, the assumption that for example an embryo or an insect is endowed with superhuman intelligence. An important part of those phenomena which have been advanced as "proofs of vitalism," such as equifinality and anamorphosis, are consequences of the characteristic state of the organism as an open system, and thus accessible to scientific interpretation and theory.

Isomorphism in Science

The present study merely intended to briefly point out the general aim and several concepts of general system theory. Further tasks on the one hand would be to express this theory in a logico-mathematically strict form; on the other hand the principles holding for any type of systems would have to be further developed. This is a concrete problem. For example, demographic dynamics may be developed homologous to mechanical dynamics (Volterra, cf. d'Ancona, 1939). A principle of minimum action may be found in various fields, in mechanics, in physical chemistry as Le Châtelier's principle which, as may be proved, is also valid for open systems, in electricity as Lenz's rule, in population theory according to Volterra, etc. A principle of relaxation oscillations occurs in physical systems as well as in many biological phenomena and certain models of population dynamics. A general theory of periodicities appears as a desideratum of various fields of science. Efforts will therefore have to be made towards a development of principles such as those of minimum action, conditions of stationary and periodic solutions (equilibria and rhythmic fluctuations), the existence of steady states and similar problems in a form generalized with respect to physics and valid for systems in general.

General system theory therefore is not a catalogue of well-known differential equations and their solutions, but raises new and well-defined problems which partly do not appear in physics, but are of basic importance in non-physical fields. Just because the phenomena concerned are not dealt with in ordinary physics, these problems have often appeared as metaphysical or vitalistic.

General system theory should further be an important regulative device in science. The existence of laws of similar structure in different fields makes possible the use of models which are simpler or better known, for more complicated and less manageable phenomena. Therefore general system theory should be, methodologically, an important means of controlling and instigating the transfer of principles from one field to another, and it will no longer be necessary to duplicate or triplicate the discovery of the same principles in different fields isolated from each other. At the same time, by formulating exact criteria, general system theory will guard against superficial analogies which are useless in science and harmful in their practical consequences.

This requires a definition of the extent to which "analogies" in science are permissible and useful.

We have previously seen the appearance of similar system laws in various sciences. The same is true of phenomena where the general principles can be described in ordinary language though they cannot be formulated in mathematical terms. For instance, there are hardly processes more unlike phenomenologically and in their intrinsic mechanisms than the formation of a whole animal out of a divided sea-urchin or newt germ, the reestablishment of normal function in the central nervous system after removal of or injury to some of its parts, and gestalt perception in psychology. Nevertheless, the principles governing these different phenomena show striking similarities. Again, when we investigate the development of the Germanic languages, it may be observed that, beginning with a primitive language, certain sound mutations occurred in parallel development in various tribes, though these were geographically located far apart from each other; in Iceland, on the British Isles, on the Iberian peninsula. Mutual influence is out of question; the languages rather developed independently after separation of the tribes, and yet show definite parallelism.* The biologist may find a corresponding principle in certain evolutionary developments. There is, for instance, the group of extinct hoofed animals, the titanotheres. During the Tertiary, they developed from smaller into gigantic forms, while with increasing body size formation of ever larger horns took place. A more detailed investigation showed that the titanotheres, starting from those small, early forms, split up into several groups which developed independently of each other but still showed parallel characteristics. Thus we find an interesting similarity in the phenomenon of parallel evolution starting from common origins but developing independently—here: the independent evolution of tribal languages; there: independent evolution of groups within a certain class of mammals.

In simple cases, the reason for isomorphism is readily seen.

*I am obliged to Prof. Otto Höfler for having indicated this phenomenon to me.

For example, the exponential law states that, given a complex of a number of entities, a constant percentage of these elements decay or multiply per unit time. Therefore this law will apply to the pounds in a banking account as well as to radium atoms, molecules, bacteria, or individuals in a population. The logistic law says that the increase, originally exponential, is limited by some restricting conditions. Thus in autocatalytic reaction, a compound catalyzes its own formation; but since the number of molecules is finite in a closed reaction vessel, the reaction must stop when all molecules are transformed, and must therefore approach a limiting value. A population increases exponentially with the increasing number of individuals, but if space and food are limited, the amount of food available per individual decreases; therefore the increase in number cannot be unlimited, but must approach a steady state defined as the maximum population compatible with resources available. Railway lines which already exist in a country lead to the intensification of traffic and industry which, in turn, make necessary a denser railway network, till a state of saturation is eventually reached; thus, railways behave like autocatalyzers accelerating their own increase, and their growth follows the autocatalytic curve. The parabolic law is an expression for competition within a system, each element taking its share according to its capacity as expressed by a specific constant. Therefore the law is of the same form whether it applies to the competition of individuals in an economic system, according to Pareto's law, or to organs competing within an organism for nutritive material and showing allometric growth.

There are obviously three prerequisites for the existence of isomorphisms in different fields and sciences. Apparently, the isomorphisms of laws rest in our cognition on the one hand, and in reality on the other. Trivially, it is easy to write down any complicated differential equation, yet even innocent-looking expressions may be hard to solve, or give, at the least, cumbersome solutions. The number of simple mathematical expressions which will be preferably applied to describe natural phenomena is limited. For this reason, laws identical in structure will appear in intrinsically different fields. The same applies to statements in ordinary language; here, too, the number of intellectual schemes is restricted, and they will be applied in quite different realms.

However, these laws and schemes would be of little help if the world (i.e., the totality of observable events) was not such that they could be applied to it. We can imagine a chaotic world or a world which is too complicated to allow the application of the relatively simple schemes which we are able to construct with our limited intellect. That this is not so is the prerequisite that science is possible. The structure of reality is such as to permit the application of our conceptual constructs. We realize, however, that all scientific laws merely represent abstractions and idealizations expressing certain aspects of reality. Every science means a schematized picture of reality, in the sense that a certain conceptual construct is unequivocally related to certain features of order in reality; just as the blueprint of a building isn't the building itself and by no means represents it in every detail such as the arrangement of bricks and the forces keeping them together, but nevertheless an unequivocal correspondence exists between the design on paper and the real construction of stone, iron and wood. The question of ultimate "truth" is not raised, that is, in how far the plan of reality as mapped by science is correct or in need or capable of improvement; likewise, whether the structure of reality is expressed in one single blueprint —i.e., the system of human science. Presumably different representations are possible or even necessary—in a similar way as it is meaningless to ask whether central or parallel projection, a horizontal or a vertical plan are more "correct." That the latter may be the case is indicated by instances where the same physical "given" can be expressed in different "languages"—e.g., by thermodynamics and statistical mechanics; or even complementary considerations become necessary, as in the corpuscle and wave models of microphysics. Independent of these questions, the existence of science proves that it is possible to express certain traits of order in reality by conceptual constructs. A presupposition for this is that order exists in reality itself; similarly—to quote the illustration mentioned—as we are able to draw the plan of a house or a crystal, but not of stones whirling around after an explosion or of the irregularly moving molecules in a liquid.

Yet there is a third reason for the isomorphism of laws in different realms which is important for the present purpose. In our considerations we started with a general definition of "system" defined as "a set of elements in interaction" and expressed by the system of equations (3.1). No special hypotheses or statements were made about the nature of the system, of its elements or the relations between them. Nevertheless from this purely formal definition of "system," many properties follow which in part are expressed in laws well-known in various fields of science, and in part concern concepts previously regarded as anthropomorphic, vitalistic or metaphysical. The parallelism of general conceptions or even special laws in different fields therefore is a consequence of the fact that these are concerned with "systems," and that certain general principles apply to systems irrespective of their nature. Hence principles such as those of wholeness and sum, mechanization, hierarchic order, approach to steady states, equifinality, etc., may appear in quite different disciplines. The isomorphism found in different realms is based on the existence of general system principles, of a more or less well-developed "general system theory."

The limitations of this conception, on the other hand, can be indicated by distinguishing three kinds or levels in the description of phenomena.

At first, there are *analogies*—i.e., superficial similarities of phenomena which correspond neither in their causal factors nor in their relevant laws. Of this kind are the *simulacra vitae*, popular in previous times, such as when the growth of an organism was compared to the growth of a crystal or of an osmotic cell. There are superficial similarities in the one or other respect, while we are safe to say that the growth of a plant or an animal does not follow the pattern of crystal growth or of an osmotic structure, and the relevant laws are different in both cases. The same applies to the consideration of a biocoenosis (e.g., a forest) as an "organism," with the obvious difference between the unification of an individual organism and the looseness of a plant association; or the comparison of the development of a population with birth, growth, aging and death of an organism where the comparison of life cycles remains highly dubious.

A second level are *homologies*. Such are present when the efficient factors are different, but the respective laws are formally identical. Such homologies are of considerable importance as conceptual models in science. They are frequently applied in physics. Examples are the consideration of heat flow as a flow of a heat substance, the comparison of electrical flow with the flow of a fluid, in general the transfer of the originally hydrodynamic notion of gradient to electrical, chemical, etc., potentials. We know exactly, of course, that there is no "heat substance" but heat is to be interpreted in the sense of kinetic theory; yet the model enables the stipulation of laws which are formally correct.

It is logical homologies with which the present investigation is concerned. We may express this as follows: If an object is a system, it must have certain general system characteristics, irrespective of what the system is otherwise. Logical homology makes possible not only isomorphy in science, but as a conceptual model has the capacity of giving instructions for correct consideration and eventual explanation of phenomena.

The third level finally is *explanation*—i.e., the statement of specific conditions and laws that are valid for an individual object or for a class of objects. In logico-mathematical language, this means that the general functions f of our equation (3.1) are replaced by specified functions applicable to the individual case. Any scientific explanation necessitates the knowledge of these specific laws as, for example, the laws of chemical equilibrium, of growth of an organism, the development of a population, etc. It is possible that also specific laws present formal correspondence or homologies in the sense discussed; but the structure of individual laws may, of course, be different in the individual cases.

Analogies are scientifically worthless. Homologies, in contrast, often present valuable models, and therefore are widely applied in physics. Similarly, general system theory can serve as a regulatory device to distinguish analogies and homologies, meaningless similarities and meaningful transfer of models. This function particularly applies to sciences which, like demography, sociology, and large fields in biology, cannot be fitted in the framework of physics and chemistry; nevertheless, there are exact laws which can be stated by application of suitable models.

The homology of system characteristics does not imply reduction of one realm to another and lower one. But neither is it mere metaphor or analogy; rather, it is a formal correspondence founded in reality inasmuch as it can be considered as constituted of "systems" of whatever kind.

Speaking philosophically, general system theory, in its developed form, would replace what is known as "theory of categories" (N. Hartmann, 1942) by an exact system of logico-mathematical laws. General notions as yet expressed in the vernacular would acquire the unambiguous and exact expression possible only in mathematical language.

The Unity of Science

We may summarize the main results of this presentation as follows:

(a) The analysis of general system principles shows that many concepts which have often been considered as anthropomorphic, metaphysical, or vitalistic are accessible to exact formulation. They are consequences of the definition of systems or of certain system conditions.

(b) Such investigation is a useful prerequisite with respect to concrete problems in science. In particular, it leads to the elucidation of problems which, in the usual schematisms and pigeon-holes of the specialized fields, are not envisaged. Thus system theory should prove an important means in the process of developing new branches of knowledge into exact science—i.e., into systems of mathematical laws.

(c) This investigation is equally important to philosophy of science, major problems of which gain new and often surprising aspects.

(d) The fact that certain principles apply to systems in general, irrespective of the nature of the systems and of the entities concerned, explains that corresponding conceptions and laws appear independently in different fields of science, causing the remarkable parallelism in their modern development. Thus, concepts such as wholeness and sum, mechanization, centralization, hierarchical order, stationary and steady states, equifinality, etc., are found in different fields of natural science, as well as in psychology and sociology.

These considerations have a definite bearing on the question of the Unity of Science. The current opinion has been well represented by Carnap (1934). As he states, Unity of Science is granted by the fact that all statements in science can ultimately be expressed in physical language—i.e., in the form of statements that attach quantitative values to definite positions in a space-time system of co-ordinates. In this sense, all seemingly nonphysical *concepts*, for instance specifically biological notions such as "species," "organism," "fertilization," and so forth, are defined by means of certain perceptible criteria—i.e., qualitative determinations capable of being physicalized. The physical language is therefore the universal language of science. The question whether biological *laws* can be reduced to physical ones—i.e., whether the natural laws sufficient to explain all inorganic phenomena are also sufficient to explain biological phenomena—is left open by Carnap, though with preference given to an answer in the affirmative.

From our point of view, Unity of Science wins a much more concrete and, at the same time, profounder aspect. We too leave open the question of the "ultimate reduction" of the laws of biology (and the other non-physical realms) to physics—i.e., the question whether a hypothetico-deductive system embracing all sciences from physics to biology and sociology may ever be established. But we are certainly able to establish scientific laws for the different levels or strata of reality. And here we find, speaking in the "formal mode" (Carnap), a correspondence or isomorphy of laws and conceptual schemes in different fields, granting the Unity of Science. Speaking in "material" language, this means that the world (i.e., the total of observable phenomena) shows a structural uniformity, manifesting itself by isomorphic traces of order in its different levels or realms.

Reality, in the modern conception, appears as a tremendous hierarchical order of organized entities, leading, in a superposition of many levels, from physical and chemical to biological and sociological systems. Unity of Science is granted, not by a utopian reduction of all sciences to physics and chemistry, but by the structural uniformities of the different levels of reality.

Especially the gap between natural and social sciences, or, to use the more expressive German terms, of *Natur- und Geisteswissenschaften*, is greatly diminished, not in the sense of a reduction of the latter to biological conceptions but in the sense of structural similarities. This is the cause of the appearance of corresponding general viewpoints and notions in both fields, and may eventually lead to the establishment of a system of laws in the latter.

The mechanistic world-view found its ideal in the Laplacean spirit—i.e., in the conception that all phenomena are ultimately aggregates of fortuitous actions of elementary physical units. Theoretically, this conception did not lead to exact sciences outside the field of physics—i.e., to laws of the higher levels of reality, the biological, psychological and sociological. Practically, its consequences have been fatal for our civilization. The attitude that considers physical phenomena as the sole standard of reality has lead to the mechanization of mankind and to the devaluation of higher values. The unregulated domination of physical technology finally ushered the world into the catastrophical crises of our time. After having overthrown the mechanistic view, we are careful not to slide into "biologism," that is, into considering mental, sociological and cultural phenomena from a merely biological standpoint. As physicalism considered the living organism as a strange combination of physico-chemical events or machines, biologism considers man as a curious zoological species, human society as a beehive or stud-farm. Biologism has, theoretically, not proved its theoretical merits, and has proved fatal in its practical consequences. The organismic conception does not mean a unilateral dominance of biological conceptions. When emphasizing general structural isomorphies of different levels, it asserts, at the same time, their autonomy and possession of specific laws.

We believe that the future elaboration of general system theory will prove to be a major step towards unification of science. It may be destined in the science of the future, to play a role similar to that of Aristotelian logic in the science of antiquity. The Greek conception of the world was static, things being considered to be a mirroring of eternal archetypes or ideas. Therefore classification was the central problem in science, the fundamental organon of which is the definition of subordination and superordination of concepts. In modern science, dynamic interaction appears to be the central problem in all fields of reality. Its general principles are to be defined by system theory.

For "Notes on Developments in Mathematical System Theory (1971)" see Appendix I, page 251.

4 Advances in General System Theory

Approaches and Aims in Systems Science

When, some 40 years ago, I started my life as a scientist, biology was involved in the mechanism-vitalism controversy. The mechanistic procedure essentially was to resolve the living organism into parts and partial processes: the organism was an aggregate of cells, the cell one of colloids and organic molecules, behavior a sum of unconditional and conditioned reflexes, and so forth. The problems of organization of these parts in the service of maintenance of the organism, of regulation after disturbances and the like were either by-passed or, according to the theory known as vitalism, explainable only by the action of soul-like factors—little hobgoblins as it were—hovering in the cell or the organism—which obviously was nothing less than a declaration of the bankruptcy of science. In this situation, I and others were led to the so-called organismic viewpoint. In one brief sentence, it means that organisms are organized things and, as biologists, we have to find out about it. I tried to implement this organismic program in various studies on metabolism, growth, and the biophysics of the organism. One step in this direction was the so-called theory of open systems and steady states which essentially is an expansion of conventional physical chemistry, kinetics and thermodynamics. It appeared, however, that I could not stop on the way once taken and so I was led to a still further generalization which I called "General System Theory." The idea goes back some considerable time: I presented it first in 1937 in Charles Morris' philosophy seminar at the University of Chicago. However, at that time theory was in bad repute in biology, and I was afraid of what Gauss, the mathematician, called the "clamor of the Boeotians." So I left my drafts in the drawer, and it was only after the war that my first publications on the subject appeared.

Then, however, something interesting and surprising happened. It turned out that a change in intellectual climate had taken place, making model building and abstract generalizations fashionable. Even more: quite a number of scientists had followed similar lines of thought. So general system theory, after all, was not isolated, not a personal idiosyncrasy as I had believed, but corresponded to a trend in modern thinking.

There are quite a number of novel developments intended to meet the needs of a general theory of systems. We may enumerate them in a brief survey:

(1) Cybernetics, based upon the principle of feedback or circular causal trains providing mechanisms for goal-seeking and self-controlling behavior.

(2) Information theory, introducing the concept of information as a quantity measurable by an expression isomorphic to negative entropy in physics, and developing the principles of its transmission.

(3) Game theory analyzing, in a novel mathematical framework, rational competition between two or more antagonists for maximum gain and minimum loss.

(4) Decision theory, similarly analyzing rational choices, within human organizations, based upon examination of a given situation and its possible outcomes.

(5) Topology or relational mathematics, including non-metrical fields such as network and graph theory.

(6) Factor analysis, i.e., isolation, by way of mathematical analysis, of factors in multivariable phenomena in psychology and other fields.

(7) General system theory in the narrower sense (G.S.T.), trying to derive, from a general definition of "system" as a complex of interacting components, concepts characteristic of organized wholes such as interaction, sum, mechanization, centralization, competition, finality, etc., and to apply them to concrete phenomena.

While systems theory in the broad sense has the character of a basic science, it has its correlate in applied science, sometimes subsumed under the general name of Systems Science. This development is closely connected with modern automation. Broadly speaking, the following fields can be distinguished (Ackoff, 1960; A.D. Hall, 1962):

Systems Engineering, i.e., scientific planning, design, evaluation, and construction of man-machine systems;

Operations research, i.e., scientific control of existing systems of men, machines, materials, money, etc.;

Human Engineering, i.e., scientific adaptation of systems and especially machines in order to obtain maximum efficiency with minimum cost in money and other expenses.

A very simple example for the necessity of study of "man-machine systems" is air travel. Anybody crossing continents by jet with incredible speed and having to spend endless hours waiting, queuing, being herded in airports, can easily realize that the physical techniques in air travel are at their best, while "organizational" techniques still are on a most primitive level.

Although there is considerable overlapping, different conceptual tools are predominant in the individual fields. In systems engineering, cybernetics and information theory and also general system theory in the narrower sense are used. Operations research uses tools such as linear programming and game theory. Human engineering, concerned with the abilities, physiological limitations and variabilities of human beings, includes biomechanics, engineering psychology, human factors, etc., among its tools.

The present survey is not concerned with applied systems science; the reader is referred to Hall's book as an excellent textbook of systems engineering (1962). However it is well to keep in mind that the systems approach as a novel concept in science has a close parallel in technology.

The motives leading to the postulate of a general theory of systems can be summarized under a few headings.

(1) Up to recent times the field of science as a nomothetic en-

deavor, i.e., trying to establish an explanatory and predictive
system of laws, was practically identical with theoretical physics.
Consequently, physical reality appeared to be the only one vouch-
safed by science. The consequence was the postulate of reductionism, i.e. the principle that biology, behavior and the social
sciences are to be handled according to the paragon of physics,
and eventually should be reduced to concepts and entities of
the physical level. Owing to developments in physics itself, the
physicalistic and reductionist theses became problematic, and
indeed appeared as metaphysical prejudices. The entities about
which physics is talking—atoms, elementary particles and the like
—have turned out to be much more ambiguous than previously
supposed: not metaphysical building blocks of the universe, but
rather complicated conceptual models invented to take account
of certain phenomena of observation. On the other hand, the
biological, behavioral and social sciences have come into their
own. Owing to the concern with these fields on the one hand, and
the exigencies of a new technology, a *generalization of scientific
concepts* and models became necessary which resulted in the
emergence of new fields beyond the traditional system of physics.

(2) In the biological, behavioral and sociological fields, there
exist predominant problems which were neglected in classical
science or rather which did not enter its considerations. If we
look at a living organism, we observe an amazing order, organization, maintenance in continuous change, regulation and apparent
teleology. Similarly, in human behavior goal-seeking and purposiveness cannot be overlooked, even if we accept a strictly behavioristic standpoint. However, concepts like organization,
directiveness, teleology, etc., just do not appear in the classic
system of science. As a matter of fact, in the so-called mechanistic
world view based upon classical physics, they were considered as
illusory or metaphysical. This means, for example, to the biologist
that just the specific problems of living nature appeared to lie
beyond the legitimate field of science. The appearance of
models—conceptual and in some cases even material—representing such aspects of multivariable interaction, organization, self-
maintenance, directiveness, etc., implies *introduction of new
categories* in scientific thought and research.

(3) Classical science was essentially concerned with two-variable
problems, one-way causal trains, one cause and one effect, or with

few variables at the most. The classical example is mechanics. It gives perfect solutions for the attraction between two celestial bodies, a sun and a planet, and hence permits exact prediction of future constellations and even the existence of still undetected planets. However, already the three-body problem of mechanics does not allow a closed solution by the analytical methods of mechanics and can only be approached by approximations. A similar situation exists in the more modern field of atomic physics (Zacharias, 1957). Here also two-body problems such as that of one proton and electron are solvable, but trouble arises with the many-body problem. Many problems, particularly in biology and the behavioral and social sciences, are essentially multivariable problems for which new conceptual tools are needed. Warren Weaver (1948), co-founder of information theory, has expressed this in an often-quoted statement. Classical science, he stated, was concerned either with linear causal trains, that is, two-variable problems; or else with unorganized complexity. The latter can be handled with statistical methods and ultimately stems from the second principle of thermodynamics. However, in modern physics and biology, *problems of organized complexity*, i.e., interaction of a large but not infinite number of variables, are popping up everywhere and demand new conceptual tools.

(4) What has been said are not metaphysical or philosophic contentions. We are not erecting a barrier between inorganic and living nature which obviously would be inappropriate in view of intermediates such as viruses, nucleoproteins and self-duplicating units. Nor do we protest that biology is in principle "irreducible to physics" which also would be out of place in view of the tremendous advances of physical and chemical explanation of life processes. Similarly, no barrier between biology and the behavioral and social sciences is intended. This, however, does not obviate the fact that in the fields mentioned we do not have appropriate conceptual tools serving for explanation and prediction as we have in physics and its various fields of application.

(5) It therefore appears that an expansion of science is required to deal with those aspects which are left out in physics and happen to concern the specific characteristics of biological, behavioral, and social phenomena. This amounts to *new conceptual models* to be introduced.

(6) These expanded and generalized theoretical constructs or

models are *interdisciplinary*—i.e., they transcend the conventional departments of science, and are applicable to phenomena in various fields. This results in the isomorphism of models, general principles and even special laws appearing in various fields.

In summary: Inclusion of the biological, behavioral and social sciences and modern technology necessitates generalization of basic concepts in science; this implies new categories of scientific thinking compared to those in traditional physics; and models introduced for such purpose are of an interdisciplinary nature.

An important consideration is that the various approaches enumerated are not, and should not be considered to be monopolistic. One of the important aspects of the modern changes in scientific thought is that there is no unique and all-embracing "world system." All scientific constructs are models representing certain aspects or perspectives of reality. This even applies to theoretical physics: far from being a metaphysical presentation of ultimate reality (as the materialism of the past proclaimed and modern positivism still implies), it is but one of these models and, as recent developments show, neither exhaustive nor unique. The various "systems theories" also are models that mirror different aspects. They are not mutually exclusive and are often combined in application. For example, certain phenomena may be amenable to scientific exploration by way of cybernetics, others by way of general system theory in the narrower sense; or even in the same phenomenon, certain aspects may be describable in the one or the other way. This, of course, does not preclude but rather implies the hope for further synthesis in which the various approaches of the present toward a theory of "wholeness" and "organization" may be integrated and unified. Actually, such further syntheses, e.g., between irreversible thermodynamics and information theory, are slowly developing.

Methods in General Systems Research

Ashby (1958a) has admirably outlined two possible ways or general methods in systems study:

Two main lines are readily distinguished. One, already well developed in the hands of von Bertalanffy and his co-workers, takes the world as we find it, examines the various systems that occur in it—zoological, physiological, and so on—and then draws up statements about the regularities that have been observed to hold. This method is essentially empirical. The second method is to start at the other end. Instead of studying first one system, then a second, then a third, and so on, it goes to the other extreme, considers the set of all conceivable systems and then reduces the set to a more reasonable size. This is the method I have recently followed.

It will easily be seen that all systems studies follow one or the other of these methods or a combination of both. Each of the approaches has its advantages as well as shortcomings.

(1) The first method is empirico-intuitive; it has the advantage that it remains rather close to reality and can easily be illustrated and even verified by examples taken from the individual fields of science. On the other hand, the approach lacks mathematical elegance and deductive strength and, to the mathematically minded, will appear naive and unsystematic.

Nevertheless, the merits of this empirico-intuitive procedure should not be minimized.

The present writer has stated a number of "system principles," partly in the context of biological theory and without explicit reference to G.S.T. (von Bertalanffy, 1960a, pp. 37–54), partly in what emphatically was entitled an "Outline" of this theory (Chapter 3). This was meant in the literal sense: It was intended to call attention to the desirability of such a field, and the presentation was in the way of a sketch or blueprint, illustrating the approach by simple examples.

However, it turned out that this intuitive survey appears to be remarkably complete. The main principles offered such as wholeness, sum, centralization, differentiation, leading part, closed and open system, finality, equifinality, growth in time, relative growth, competition, have been used in manifold ways (e.g., general definition of system: Hall and Fagen, 1956; types of growth: Keiter, 1951–52; systems engineering: A.D. Hall, 1962; social work: Hearn, 1958). Excepting minor variations in terminology intended for clarification or due to the subject matter, no principles of similar significance were added—even though this would be highly desirable. It is perhaps even more significant that this also applies to considerations which do not refer to the present writer's work and hence cannot be said to be unduly influenced by it. Perusal of studies such as those by Beer (1960) and Kremyanskiy (1960) on principles, Bradley and Calvin (1956) on the network of chemical reactions, Haire (1959) on growth of organizations, etc., will easily show that they are also using the "Bertalanffy principles."

(2) The way of deductive systems theory was followed by Ashby (1958b). A more informal presentation which summarizes Ashby's reasoning (1962) lends itself particularly well to analysis.

Ashby asks about the "fundamental concept of machine" and answers the question by stating "that its internal state, and the state of its surroundings, defines uniquely the next state it will go to." If the variables are continuous, this definition corresponds to the description of a dynamic system by a set of ordinary differential equations with time as the independent variable. However, such representation by differential equations is too restricted for a theory to include biological systems and calculating machines where discontinuities are ubiquitous. Therefore the modern definition is the "machine with input": It is defined by a set S of internal states, a set I of input and a mapping f of the product set $I \times S$ into S. "Organization," then, is defined by specifying the machine's states S and its conditions I. If S is a product set $S = \pi_i T_i$, with i as the parts and T specified by the mapping f, a "self-organizing" system, according to Ashby, can have two meanings, namely: (1) The system starts with its parts separate, and these parts then change toward forming connections (example: cells of the embryo, first having little or no effect on one another, join by formation of dendrites and synapses to form the highly interdependent nervous system). This first meaning is "changing from unorganized to organized." (2) The second meaning is "changing from a bad organization to a good one" (examples: a child whose brain organization makes it fire-seeking at first, while a new brain organization makes him fire-avoiding; an automatic pilot and plane coupled first by deleterious positive feedback and then improved). "There the organization is bad. The system would be 'self-organizing' if a change were automatically made" (changing positive into negative feedback). But "*no machine can be self-organizing in this sense*" (author's italics). For adaptation (e.g., of the homeostat or in a self-programming computer) means that we start with a set S of states, and that f changes into g, so that organization is a variable, e.g., a function of time a (t) which has first the value f and later the value g.

However, this change "cannot be ascribed to any cause in the set *S; so it must come from some outside agent, acting on the system S as input*" (our italics). In other terms, to be "self-organizing" the machine *S* must be coupled to another machine.

This concise statement permits observation of the limitations of this approach. We completely agree that description by differential equations is not only a clumsy but, in principle, inadequate way to deal with many problems of organization. The author was well aware of this, emphasizing that a system of simultaneous differential equations is by no means the most general formulation and is chosen only for illustrative purposes (Chapter 3).

However, in overcoming this limitation, Ashby introduced another one. His "modern definition" of system as a "machine with input" as reproduced above, supplants the general system model by another rather special one: the cybernetic model—i.e., a system open to information but closed with respect to entropy transfer. This becomes apparent when the definition is applied to "self-organizing systems." Characteristically, the most important kind of these has no place in Ashby's model, namely systems organizing themselves by way of progressive differentiation, evolving from states of lower to states of higher complexity. This is, of course, the most obvious form of "self-organization," apparent in ontogenesis, probable in phylogenesis, and certainly also valid in many social organizations. We have here not a question of "good" (i.e., useful, adaptive) or "bad" organization which, as Ashby correctly emphasizes, is relative to circumstances; increase in differentiation and complexity—whether useful or not—is a criterion that is objective and at least in principle amenable to measurement (e.g., in terms of decreasing entropy, of information). Ashby's contention that "no machine can be self-organizing," more explicitly, that the "change cannot be ascribed to any cause in the set *S*" but "must come from some outside agent, an input" amounts to exclusion of self-differentiating systems. The reason that such systems are not permitted as "Ashby machines" is patent. Self-differentiating systems that evolve toward higher complexity (decreasing entropy) are, for thermodynamic reasons, possible only as open systems—e.g., systems importing matter containing free energy to an amount overcompensating the increase in entropy due to irreversible processes within the system ("import of negative entropy" in Schrödinger's expression). However, we cannot say that "this change comes from some outside agent, an input"; the differentiation within a developing embryo and organism is due to its internal laws of organization, and the input (e.g., oxygen supply which may vary quantitatively, or nutrition which can vary qualitatively within a broad spectrum) makes it only possible energetically.

The above is further illustrated by additional examples given by Ashby. Suppose a digital computer is carrying through multiplications at random; then the machine will "evolve" toward showing even numbers (because products even $\times$ even as well as even $\times$ odd give numbers even), and eventually only zeros will be "surviving." In still another version Ashby quotes Shannon's Tenth Theorem, stating that if a correction channel has capacity H, equivocation of the amount H can be removed, but no more. Both examples illustrate the working of closed systems: The "evolution" of the computer is one toward disappearance of differentiation and establishment of maximum homogeneity (analog to the second principle in closed systems); Shannon's Theorem similarly concerns closed systems where no negative entropy is fed in. Compared to the information content (organization) of a living system, the imported matter (nutrition, etc.) carries not information but "noise." Nevertheless, its negative entropy is used to maintain or even to increase the information content of the system. This is a state of affairs apparently not provided for in Shannon's Tenth Theorem, and understandably so as he is not treating information transfer in open systems with transformation of matter.

In both respects, the living organism (and other behavioral and social systems) is not an Ashby machine because it evolves toward increasing differentiation and inhomogeneity, and can correct "noise" to a higher degree than an inanimate communication channel. Both, however, are consequences of the organism's character as an open system.

Incidentally, it is for similar reasons that we cannot replace the concept of "system" by the generalized "machine" concept of Ashby. Even though the latter is more liberal compared to the classic one (machines defined as systems with fixed arrangement of parts and processes), the objections against a "machine theory" of life (von Bertalanffy, 1960, pp. 16–20 and elsewhere) remain valid.

These remarks are not intended as adverse criticism of Ashby's or the deductive approach in general; they only emphasize that there is no royal road to general systems theory. As every other scientific field, it will have to develop by an interplay of empirical, intuitive and deductive procedures. If the intuitive approach leaves much to be desired in logical rigor and completeness, the deductive approach faces the difficulty of whether the fundamental terms are correctly chosen. This is not a particular fault of the theory or of the workers concerned but a rather common phenomenon in the history of science; one may, for example, remember the long debate as to what magnitude—force or energy—is to be considered as constant in physical transformations until the issue was decided in favor of $mv^2/2$.

In the present writer's mind, G.S.T. was conceived as a working hypothesis; being a practicing scientist, he sees the main function of theoretical models in the explanation, prediction and control of hitherto unexplored phenomena. Others may, with equal right, emphasize the importance of axiomatic approach and quote to this effect examples like the theory of probability, non-Euclidean geometries, more recently information and game theory, which were first developed as deductive mathematical fields and later applied in physics or other sciences. There should be no quarrel about this point. The danger, in both approaches, is to consider too early the theoretical model as being closed and definitive—a danger particularly important in a field like general systems which is still groping to find its correct foundations.

Advances of General System Theory

The decisive question is that of the explanatory and predictive value of the "new theories" attacking the host of problems around wholeness, teleology, etc. Of course, the change in intellectual climate which allows one to see new problems which were overlooked previously, or to see problems in a new light, is in a way more important than any single and special application. The "Copernican Revolution" was more than the possibility somewhat better to calculate the movement of the planets; general relativity more than an explanation of a very small number of recalcitrant phenomena in physics; Darwinism more than a hypothetical answer to zoological problems; it was the changes in the general frame of reference that mattered (cf. Rapoport, 1959a). Nevertheless, the justification of such change ultimately is in specific achievements which would not have been obtained without the new theory.

There is no question that new horizons have been opened up but the relations to empirical facts often remain tenuous. Thus, information theory has been hailed as a "major breakthrough," but outside the original technological field, contributions have remained scarce. In psychology, they are so far limited to rather trivial applications such as rote learning, etc. (Rapoport, 1956, Attneave, 1959). When, in biology, DNA is spoken of as "coded information" and of "breaking the code" when the structure of nucleic acids is elucidated, use of the term information is a *façon de parler* rather than application of information theory in the technical sense as developed by Shannon and Weaver (1949). "Information theory, although useful for computer design and network analysis, has so far not found a significant place in biology" (Bell, 1962). Game theory, too, is a novel mathematical development which was considered to be comparable in scope to Newtonian mechanics and the introduction of calculus; again, "the applications are meager and faltering" (Rapoport, 1959a; the reader is urgently referred to Rapoport's discussions on information and game theory which admirably analyze the problems here mentioned). The same is seen in decision theory from which considerable gain in applied systems science was expected; but as regards the much-advertised military and business games, "there has been no controlled evaluation of their performance in training, personnel selection, and demonstration" (Ackoff, 1959).

A danger in recent developments should not remain unmentioned. Science of the past (and partly still the present) was dominated by one-sided empiricism. Only collection of data and experiments were considered as being "scientific" in biology (and psychology); "theory" was equated with "speculation" or "philosophy," forgetting that a mere accumulation of data, although steadily piling up, does not make a "science." Lack of recognition and support for development of the necessary theoretical framework and unfavorable influence on experimental research itself (which largely became an at-random, hit-or-miss endeavor) was the consequence (cf. Weiss, 1962a). This has, in certain fields, changed to the contrary in recent years. Enthusiasm for the new mathematical and logical tools available has led to feverish "model building" as a purpose in itself and often without regard to empirical fact. However, conceptual experimentation at random has no greater chances of success than at-random experimentation in the laboratory. In the words of Ackoff (1959), there is the fundamental misconception in game (and other) theory to mistake for a "problem" what actually is only a mathematical "exercise." One would do well to remember the old Kantian maxim that experience without theory is blind but theory without experience a mere intellectual play.

The case is somewhat different with cybernetics. The model here applied is not new; although the enormous development in the field dates from the introduction of the name, Cybernetics (Wiener, 1948), application of the feedback principle to physiological processes goes back to R. Wagner's work nearly 40 years ago (cf. Kment, 1959). The feedback and homeostasis model has since been applied to innumerable biological phenomena and—somewhat less persuasively—in psychology and the social sciences. The reason for the latter fact is, in Rapoport's words (1956) that

> usually, there is a well-marked correlation between the scope and the soundness of the writings. . . . The sound work is confined either to engineering or to rather trivial applications; ambitious formulations remain vague.

This, of course, is an ever-present danger in all approaches to general systems theory: doubtless, there is a new compass of thought but it is difficult to steer between the Scylla of the trivial and the Charybdis of mistaking neologisms for explanation.

The following survey is limited to "classical" general system theory—"classical" not in the sense that it claims any priority or excellence, but that the models used remain in the framework of "classical" mathematics in contradistinction to the "new" mathematics in game, network, information theory, etc. This does not imply that the theory is merely application of conventional mathematics. On the contrary, the system concept poses problems which are partly far from being answered. In the past, system problems have led to important mathematical developments such as Volterra's theory of integro-differential equations, of systems with "memory" whose behavior depends not only on actual conditions but also on previous history. Presently important problems are waiting for further developments, e.g., a general theory of non-linear differential equations, of steady states and rhythmic phenomena, a generalized principle of least action, the thermodynamic definition of steady states, etc.

It is, of course, irrelevant whether or not research was explicitly labeled as "general system theory." No complete or exhaustive review is intended. The aim of this unpretentious survey will be fulfilled if it can serve as a sort of guide to research done in the field, and to areas that are promising for future work.

Open Systems

The theory of open systems is an important generalization of physical theory, kinetics and thermodynamics. It has led to new principles and insight, such as the principle of equifinality, the generalization of the second thermodynamic principle, the possible increase of order in open systems, the occurrence of periodic phenomena, of overshoot and false start, etc.

The extensive work in biology and related fields is partly reviewed in Chapters 5–7. (For further discussion also cf. Bray and White, 1957; Jung, 1956; Morchio, 1956; Netter, 1953, 1959).

Beyond the individual organism, systems principles are also used in population dynamics and ecologic theory (review: J. R. Bray, 1958). Dynamic ecology, i.e., the succession and climax of plant populations, is a much-cultivated field which, however, shows a tendency to slide into verbalism and terminological debate. The systems approach seems to offer a new viewpoint. Whittacker (1953) has described the sequence of plant communities toward a climax formation in terms of open systems and equifinality. According to this author, the fact that similar climax formations may develop from different initial vegetations is a striking example of equifinality, and one where the degree of independence of starting conditions and the course development has taken appear even greater than in the individual organism. A quantitative analysis on the basis of open systems in terms of production of biomass, with climax as steady state attained, was given by Patten (1959).

The open-system concept has also found application in the earth sciences, geomorphology (Chorley, 1964) and meteorology (Thompson, 1961) drawing a detailed comparison of modern meteorological concepts and Bertalanffy's organismic concept in biology. It may be remembered that already Prigogine in his classic (1947) mentioned meteorology as one possible field of application of open systems.

GROWTH-IN-TIME

The simplest forms of growth which, for this reason, are particularly apt to show the isomorphism of law in different fields are the exponential and the logistic. Examples are, among many others, the increase of knowledge of number of animal species (Gessner, 1952), publications on drosophila (Hersh, 1942), of manufacturing companies (Haire, 1959). Boulding (1956a) and Keiter (1951–52) have emphasized a general theory of growth.

The theory of animal growth after Bertalanffy (and others)—which, in virtue of using overall physiological parameters ("anabolism," "catabolism") may be subsumed under the heading of G.S.T. as well as under that of biophysics—has been surveyed in its various applications (Bertalanffy, 1960b).

RELATIVE GROWTH

A principle which is also of great simplicity and generality concerns the relative growth of components within a system. The simple relationship of allometric increase applies to many growth phenomena in biology (morphology, biochemistry, physiology, evolution).

A similar relationship obtains in social phenomena. Social differentiation and division of labor in primitive societies as well as the process of urbanization (i.e., growth of cities in comparison to rural population) follow the allometric equation. Application of the latter offers a quantitative measure of social organization and development, apt to replace the usual, intuitive judgments (Naroll and Bertalanffy, 1956). The same principle apparently applies to the growth of staff compared to total number of employees in manufacturing companies (Haire, 1959).

COMPETITION AND RELATED PHENOMENA

The work in population dynamics by Volterra, Lotka, Gause and others belongs to the classics of G.S.T., having first shown that it is possible to develop conceptual models for phenomena such as the "struggle for existence" that can be submitted to empirical test. Population dynamics and related population genetics have since become important fields in biological research.

It is important to note that investigation of this kind belongs not only to basic but also to applied biology. This is true of fishery biology where theoretical models are used to establish optimum conditions for the exploitation of the sea (survey of the more important models: Watt, 1958). The most elaborate dynamic model is by Beverton and Holt (1957; short survey: Holt, w.y.) developed for fish populations exploited in commercial fishery but certainly of wider application. This model takes into account recruitment (i.e., entering of individuals into the population), growth (assumed to follow the growth equations after Bertalanffy), capture (by exploitation), and natural mortality. The practical value of this model is illustrated by the fact that it has been adopted for routine purposes by the Food and Agriculture Organization of the United Nations, the British Ministry of Agriculture and Fisheries and other official agencies.

Richardson's studies on armaments races (cf. Rapoport, 1957, 1960), notwithstanding their shortcomings, dramatically show the possible impact of the systems concept upon the most vital concerns of our time. If rational and scientific considerations matter at all, this is one way to refute such catchwords as *Si vis pacem para bellum*.

The expressions used in population dynamics and the biological "struggle for existence," in econometrics, in the study of armament races (and others) all belong to the same family of equations (the system discussed in Chapter 3). A systematic comparison and study of these parallelisms would be highly interesting and rewarding (cf. also Rapoport, 1957, p. 88). One may, for example, suspect that the laws governing business cycles and those of population fluctuations according to Volterra stem from similar conditions of competition and interaction in the system.

In a non-mathematical way, Boulding (1953) has discussed what he calls the "Iron Laws" of social organizations: the Malthusian law, the law of optimum size of organizations, existence of cycles, the law of oligopoly, etc.

Systems Engineering

The theoretical interest of systems engineering and operations research is in the fact that entities whose components are most heterogeneous—men, machines, buildings, monetary and other values, inflow of raw material, outflow of products and many other items—can successfully be submitted to systems analysis.

As already mentioned, systems engineering employs the methodology of cybernetics, information theory, network analysis, flow and block diagrams, etc. Considerations of G.S.T. also enter (A.D. Hall, 1962). The first approaches are concerned with structured, machine-like aspects (yes-or-no decisions in the case of information theory); one would suspect that G.S.T. aspects will win increased importance with dynamic aspects, flexible organizations, etc.

PERSONALITY THEORY

Although there is an enormous amount of theorizing on neural and psychological function in the cybernetic line based upon the brain-computer comparison, few attempts have been made to apply G.S.T. in the narrower sense to the theory of human behavior (e.g., Krech, 1956; Menninger, 1957). For the present purposes, the latter may be nearly equated with personality theory.

We have to realize at the start that personality theory is at present a battlefield of contrasting and controversial theories. Hall and Lindzey (1957, p. 71) have justly stated: "All theories of behavior are pretty poor theories and all of them leave much to be desired in the way of scientific proof"—this being said in a textbook of nearly 600 pages on "Theories of Personality."

We can therefore not well expect that G.S.T. can present solutions where personality theorists from Freud and Jung to a host of modern writers were unable to do so. The theory will have shown its value if it opens new perspectives and viewpoints capable of experimental and practical application. This appears to be the case. There is quite a group of psychologists who are committed to an organismic theory of personality, Goldstein and Maslow being well-known representatives.

There is, of course, the fundamental question whether, first, G.S.T. is not essentially a physicalistic simile, inapplicable to psychic phenomena; and secondly whether such model has explanatory value when the pertinent variables cannot be defined quantitatively as is in general the case with psychological phenomena.

(1) The answer to the first question appears to be that the

system concept is abstract and general enough to permit application to entities of whatever denomination. The notions of "equilibrium," "homeostasis," "feedback," "stress," etc., are no less of technologic or physiological origin but more or less successfully applied to psychological phenomena. System theorists agree that the concept of "system" is not limited to material entities but can be applied to any "whole" consisting of interacting "components."

(2) If quantization is impossible, and even if the components of a system are ill-defined, it can at least be expected that certain principles will qualitatively apply to the whole *qua* system. At least "explanation in principle" (see below) may be possible.

Bearing in mind these limitations, one concept which may prove to be of a key nature is the organismic notion of the organism as a spontaneously active system. In the present author's words,

> Even under constant external conditions and in the absence of external stimuli the organism is not a passive but a basically active system. This applies in particular to the function of the nervous system and to behavior. It appears that internal activity rather than reaction to stimuli is fundamental. This can be shown with respect both to evolution in lower animals and to development, for example, in the first movements of embryos and fetuses (von Bertalanffy, 1960a).

This agrees with what von Holst has called the "new conception" of the nervous system, based upon the fact that primitive locomotor activities are caused by central automatisms that do not need external stimuli. Therefore, such movements persist, for example, even after the connection of motoric to sensory nerves had been severed. Hence the reflex in the classic sense is not the basic unit of behavior but rather a regulatory mechanism superposed upon primitive, automatic activities. A similar concept is basic in the theory of instinct. According to Lorenz, innate releasing mechanisms (I.R.M.) play a dominant role, which sometimes go off without an external stimulus (in vacuo or running-idle reactions): A bird which has no material to build a nest may perform the movements of nest building in the air. These considerations are in the framework of what Hebb (1955) called the "conceptual C.N.S. of 1930–1950." The more recent insight into activating systems of the brain emphasizes differently, and with a wealth of experimental evidence, the same basic concept of the autonomous activity of the C.N.S.

The significance of these concepts becomes apparent when we consider that they are in fundamental contrast to the conventional stimulus-response scheme which assumes that the organism is an essentially reactive system answering, like an automaton, to external stimuli. The dominance of the S-R scheme in contemporary psychology needs no emphasis, and is obviously connected with the *zeitgeist* of a highly mechanized society. This principle is basic in psychological theories which in all other respects are opposite, for example, in behavioristic psychology as well as in psychoanalysis. According to Freud it is the supreme tendency of the organism to get rid of tensions and drives and come to rest in a state of equilibrium governed by the "principle of stability" which Freud borrowed from the German philosopher, Fechner. Neurotic and psychotic behavior, then, is a more or less effective or abortive defense mechanism tending to restore some sort of equilibrium (according to D. Rapaport's analysis (1960) of the structure of psychoanalytic theory: "economic" and "adaptive points of view").

Charlotte Bühler (1959), the well-known child psychologist, has aptly epitomized the theoretical situation:

> In the fundamental psychoanalytic model, there is only one basic tendency, that is toward *need gratification* or *tension reduction*. . . . Present-day biologic theories emphasize the "spontaneity" of the organism's activity which is due to its built-in energy. The organism's autonomous functioning, its "drive to perform certain movements" is emphasized by Bertalanffy. . . . These concepts represent *a complete revision of the original homeostasis principle* which emphasized exclusively the tendency toward equilibrium. It is the original homeostasis principle with which psychoanalysis identified its theory of discharge of tensions as the only primary tendency (italics partly ours).

In brief, we may define our viewpoint as "Beyond the Homeostatic Principle":

(1) The S-R scheme misses the realms of play, exploratory activities, creativity, self-realization, etc.;

(2) The economic scheme misses just specific, human achievements—the most of what loosely is termed "human culture";

(3) The equilibrium principle misses the fact that psychological and behavioral activities are more than relaxation of tensions; far from establishing an optimal state, the latter may entail psychosis-like disturbances as, e.g., in sensory-deprivation experiments.

It appears that the S-R and psychoanalytic model is a highly unrealistic picture of human nature and, in its consequences, a rather dangerous one. Just what we consider to be specific human achievements can hardly be brought under the utilitarian, homeostasis, and stimulus-response scheme. One may call mountain climbing, composing of sonatas or lyrical poems "psychological homeostasis"—as has been done—but at the risk that this physiologically well-defined concept loses all meaning. Furthermore, if the principle of homeostatic maintenance is taken as a golden rule of behavior, the so-called well-adjusted individual will be the ultimate goal, that is a well-oiled robot maintaining itself in optimal biological, psychological and social homeostasis. This is *Brave New World*—not, for some at least, the ideal state of humanity. Furthermore, that precarious mental equilibrium must not be disturbed: Hence, in what rather ironically is called progressive education, the anxiety not to overload the child, not to impose constraints and to minimize all directing influences—with the result of a previously unheard-of crop of illiterates and juvenile delinquents.

In contrast to conventional theory, it can safely be maintained that not only stresses and tensions but equally complete release from stimuli and the consequent mental void may be neurosogenic or even psychosogenic. Experimentally this is verified by the experiments with sensory deprivation when subjects, insulated from all incoming stimuli, after a few hours develop a so-called model psychosis with hallucinations, unbearable anxiety, etc. Clinically it amounts to the same when insulation leads to prisoners' psychosis and to exacerbation of mental disease by isolation of patients in the ward. In contrast, maximal stress need not necessarily produce mental disturbance. If conventional theory were correct, Europe during and after the war, with extreme physiological as well as psychological stresses, should have been a gigantic lunatic asylum. As a matter of fact, there was statistically no increase either in neurotic or psychotic disturbances, apart from easily explained acute disturbances such as combat neurosis (see Chapter 9).

So we arrive at the conception that a great deal of biological and human behavior is beyond the principles of utility, homeostasis and stimulus-response, and that it is just this which is characteristic of human and cultural activities. Such a new look opens new perspectives not only in theory but in practical implications with respect to mental hygiene, education, and society in general. (See Chapter 9).

What has been said can also be couched in philosophical terms. If existentialists speak of the emptiness and meaninglessness of life, if they see in it a source not only of anxiety but of actual mental illness, it is essentially the same viewpoint: that behavior is not merely a matter of satisfaction of biological drives and of maintenance in psychological and social equilibrium but that something more is involved. If life becomes unbearably empty in an industrialized society, what can a person do but develop a neurosis? The principle, which may loosely be called spontaneous activity of the psychophysical organism, is a more realistic formulation of what the existentialists want to say in their often obscure language. And if personality theorists like Maslow or Gardner Murphy speak of self-realization as human goal, it is again a somewhat pompous expression of the same.

THEORETICAL HISTORY

We eventually come to those highest and ill-defined entities that are called human cultures and civilizations. It is the field often called "philosophy of history." We may perhaps better speak of "theoretical history," admittedly in its very first beginnings. This name expresses the aim to form a connecting link between "science" and the "humanities"; more in particular, between the social sciences and history.

It is understood, of course, that the techniques in sociology and history are entirely different (polls, statistical analysis against archival studies, internal evidence of historic relics, etc.). However, the object of study is essentially the same. Sociology is essentially concerned with a temporal cross-section as human societies *are*; history with the "longitudinal" study how societies *become* and develop. The object and techniques of study certainly justify practical differentiation; it is less clear, however, that they justify fundamentally different philosophies.

The last statement already implies the question of constructs in history, as they were presented, in grand form, from Vico to Hegel, Marx, Spengler, and Toynbee. Professional historians regard them at best as poetry, at worst as fantasies pressing—with paranoic obsession—the facts of history into a theoretical bed of Procrustes. It seems history can learn from the system theorists not ultimate solutions but a sounder methodological outlook. Problems hitherto considered to be philosophical or metaphysical can well be defined in their scientific meaning, with some interesting outlook at recent developments (e.g., game theory) thrown into the bargain.

Empirical criticism is outside the scope of the present study. For example, Geyl (1958) and many others have analyzed obvious misrepresentations of historical events in Toynbee's work, and even the non-specialist reader can easily draw a list of fallacies especially in the later, Holy Ghost–inspired volumes of Toynbee's *magnum opus*. The problem, however, is larger than errors in fact or interpretation or even the question of the merits of Marx's, Spengler's or Toynbee's theories; it is whether, in principle, models and laws are admissible in history.

A widely held contention says that they are not. This is the concept of "nomothetic" method in science and "idiographic" method in history. While science to a greater or less extent can establish "laws" for natural events, history, concerned with human events of enormous complexity in causes and outcome and presumably determined by free decisions of individuals, can only describe, more or less satisfactorily, what has happened in the past.

Here the methodologist has his first comment. In the attitude just outlined, academic history condemns constructs of history as "intuitive," "contrary to fact," "arbitrary," etc. And, no doubt, the criticism is pungent enough vis-à-vis Spengler or Toynbee. It is, however, somewhat less convincing if we look at the work of conventional historiography. For example, the Dutch historian, Peter Geyl, who made a strong argument against Toynbee from such methodological considerations, also wrote a brilliant book about Napoleon (1949), amounting to the result that there are a dozen or so different interpretations—we may safely say, *models*

—of Napoleon's character and career within academic history, all based upon "fact" (the Napoleonic period happens to be one of the best documented) and all flatly contradicting each other. Roughly speaking, they range from Napoleon as the brutal tyrant and egotistic enemy of human freedom to Napoleon the wise planner of a unified Europe; and if one is a Napoleonic student (as the present writer happens to be in a small way), one can easily produce some original documents refuting misconceptions occurring even in generally accepted, standard histories. You cannot have it both ways. If even a figure like Napoleon, not very remote in time and with the best of historical documentation, can be interpreted contrarily, you cannot well blame the "philosophers of history" for their intuitive procedure, subjective bias, etc., when they deal with the enormous phenomenon of universal history. What you have in both cases is a conceptual model which always will represent certain aspects only, and for this reason will be one-sided or even lopsided. Hence the construction of conceptual models in history is not only permissible but, as a matter of fact, is at the basis of any historical interpretation as distinguished from mere enumeration of data—i.e., chronicle or annals.

If this is granted, the antithesis between idiographic and nomothetic procedure reduces to what psychologists are wont to call the "molecular" and "molar" approach. One can analyze events within a complex whole—individual chemical reactions in an organism, perceptions in the psyche, for example; or one can look for overall laws covering the whole such as growth and development in the first or personality in the second instance. In terms of history, this means detailed study of individuals, treaties, works of art, singular causes and effects, etc., or else overall phenomena with the hope of detecting grand laws. There are, of course, all transitions between the first and second considerations; the extremes may be illustrated by Carlyle and his hero worship at one pole and Tolstoy (a far greater "theoretical historian" than commonly admitted) at the other.

The question of a "theoretical history" therefore is essentially that of "molar" models in the field; and this is what the constructs of history amount to when divested of their philosophical embroidery.

The evaluation of such models must follow the general rules

for verification or falsification. First, there is the consideration of empirical bases. In this particular instance, it amounts to the question whether or not a limited number of civilizations—some 20 at the best—provide a sufficient and representative sample to establish justified generalizations. This question and that of the value of proposed models will be answered by the general criterion: whether or not the model has explanatory and predictive value, i.e., throws new light upon known facts and correctly foretells facts of the past or future not previously known.

Although elementary, these considerations nevertheless are apt to remove much misunderstanding and philosophical fog which has clouded the issue.

(1) As has been emphasized, the evaluation of models should be simply pragmatic in terms of their explanatory and predictive merits (or lack thereof); *a priori* considerations as to their desirability or moral consequences do not enter.

Here we encounter a somewhat unique situation. There is little objection against so-called "synchronic" laws—i.e., supposed regularities governing societies at a certain point in time; as a matter of fact, beside empirical study this is the aim of sociology. Also certain "diachronic" laws—i.e., regularities of development in time—are undisputed such as, e.g., Grimm's law stating rules for the changes of consonants in the evolution of Indo-Germanic languages. It is commonplace that there is a sort of "life cycle" —stages of primitivity, maturity, baroque dissolution of form and eventual decay for which no particular external causes can be indicated—in individual fields of culture, such as Greek sculpture, Renaissance painting or German music. Indeed, this even has its counterpart in certain phenomena of biological evolution showing, as in ammonites or dinosaurs, a first explosive phase of formation of new types, followed by a phase of speciation and eventually of decadence.

Violent criticism comes in when this model is applied to civilization as a whole. It is a legitimate question—why often rather unrealistic models in the social sciences remain matters of academic discussion, while models of history encounter passionate resistance? Granting all factual criticism raised against Spengler or Toynbee, it seems rather obvious that emotional factors are involved. The highway of science is strewn with corpses of deceased theories which just decay or are preserved as mummies in the museum of history of science. In contrast, historical constructs and especially theories of historical cycles appear to touch a raw nerve, and so opposition is much more than usual criticism of a scientific theory.

(2) This emotional involvement is connected with the question of "Historical Inevitability" and a supposed degradation of human "freedom." Before turning to it, discussion of mathematical and non-mathematical models is in order.

Advantages and shortcomings of mathematical models in the social sciences are well known (Arrow, 1956; Rapoport, 1957). Every mathematical model is an oversimplification, and it remains questionable whether it strips actual events to the bones or cuts away vital parts of their anatomy. On the other hand, so far as it goes, it permits necessary deduction with often unexpected results which would not be obtained by ordinary "common sense."

In particular, Rashevsky has shown in several studies how mathematical models of historical processes can be constructed (Rashevsky, 1951, 1952).

On the other hand, the value of purely qualitative models should not be underestimated. For example, the concept of "ecologic equilibrium" was developed long before Volterra and others introduced mathematical models; the theory of selection belongs to the stock-in-trade of biology, but the mathematical theory of the "struggle for existence" is comparatively recent, and far from being verified under wildlife conditions.

In complex phenomena, "explanation in principle" (Hayek, 1955) by qualitative models is preferable to no explanation at all. This is by no means limited to the social sciences and history; it applies alike to fields like meteorology or evolution.

(3) "Historical inevitability"—subject of a well-known study by Sir Isaiah Berlin (1955) —dreaded as a consequence of "theoretical history," supposedly contradicting our direct experience of having free choices and eliminating all moral judgment and values—is a phantasmagoria based upon a world view which does not exist any more. As in fact Berlin emphasizes, it is founded upon the concept of the Laplacean spirit who is able completely to predict the future from the past by means of deterministic laws. This has no resemblance to the modern concept of "laws of nature." All "laws of nature" have a statistical character. They do not predict

an inexorably determined future but probabilities which, depending on the nature of events and on the laws available, may approach certainty or else remain far below it. It is nonsensical to ask for or fear more "inevitability" in historical theory than is found in sciences with relatively high sophistication like meteorology or economics.

Paradoxically, while the cause of free will rests with the testimony of intuition or rather immediate experience and can never be proved objectively ("Was it Napoleon's free will that led him to the Russian Campaign?"), determinism (in the statistical sense) can be proved, at least in small-scale models. Certainly business depends on personal "initiative," the individual "decision" and "responsibility" of the entrepreneur; the manager's choice whether or not to expand business by employing new appointees is "free" in precisely the sense as Napoleon's choice of whether or not to accept battle at the Moskwa. However, when the growth curve of industrial companies is analyzed, it is found that "arbitrary" deviations are followed by speedy return to the normal curve, as if invisible forces were active. Haire (1959, p. 283) states that "the return to the pattern predicted by earlier growth suggests the operation of *inexorable forces* operating on the social organism" (our italics).

It is characteristic that one of Berlin's points is "the fallacy of historical determinism (appearing) from its utter inconsistency with the common sense and everyday life of looking at human affairs." This characteristic argument is of the same nature as the advice not to adopt the Copernican system because everybody can see that the sun and not the earth moves from morning to evening.

(4) Recent developments in mathematics even allow to submit "free will"—apparently the philosophical problem most resistant to scientific analysis—to mathematical examination.

In the light of modern systems theory, the alternative between molar and molecular, nomothetic and idiographic approach can be given a precise meaning. For mass behavior, system laws would apply which, if they can be mathematized, would take the form of differential equations of the sort of those used by Richardson (cf. Rapoport, 1957) mentioned above. In contrast, free choice of the individual would be described by formulations of the nature of game and decision theory.

Axiomatically, game and decision theory are concerned with "rational" choice. This means a choice which "maximizes the individual's utility or satisfaction," that "the individual is free to choose among several possible courses of action and decides among them at the basis of their consequences," that he "selects, being informed of all conceivable consequences of his actions, what stands highest on his list," he "prefers more of a commodity to less, other things being equal," etc. (Arrow, 1956). Instead of economical gain, any higher value may be inserted without changing the mathematical formalism.

The above definition of "rational choice" includes everything that can be meant by "free will." If we do not wish to equate "free will" with complete arbitrariness, lack of any value judgment and therefore completely inconsequential actions (like the philosopher's favorite example: It is my free will whether or not to wiggle my left little finger), it is a fair definition of those actions with which the moralist, priest or historian is concerned: free decision between alternatives based upon insight into the situation and its consequences and guided by values.

The difficulty to apply theory even to simple, actual situations is of course enormous; so is the difficulty in establishing overall laws. However, without explicit formulation, both approaches can be evaluated in principle—leading to an unexpected paradox.

The "principle of rationality" fits—not the majority of human actions but rather the "unreasoning" behavior of animals. Animals and organisms in general do function in a "ratiomorphic" way, maximizing such values as maintenance, satisfaction, survival, etc.; they select, in general, what is biologically good for them, and prefer more of a commodity (e.g., food) to less.

Human behavior, on the other hand, falls far short of the principle of rationality. It is not even necessary to quote Freud to show how small is the compass of rational behavior in man. Women in a supermarket, in general, do not maximize utility but are susceptible to the tricks of the advertiser and packer; they do not make a rational choice surveying all possibilities and consequences; and do not even prefer more of the commodity packed in an inconspicuous way to less when packed in a big red box with attractive design. In our society, it is the job of an influential specialty—advertisers, motivation researchers, etc.—to *make* choices irrational which essentially is done by coupling biological factors—conditioned reflex, unconscious drives—with symbolic values (cf. von Bertalanffy, 1956a).

And there is no refuge by saying that this irrationality of human behavior concerns only trivial actions of daily life; the same principle applies to "historical" decisions. That wise old bird Oxenstierna, Sweden's Chancellor during the Thirty Years' War, expressed this perfectly by saying: *Nescis, mi fili, quantilla ratione mundus regatur*—you don't know, my dear boy, with what little reason the world is governed. Reading newspapers or listening to the radio readily shows that this applies perhaps even more to the 20th than the 17th century.

Methodologically, this leads to a remarkable conclusion. If one of the two models is to be applied, and if the "actuality principle" basic in historical fields like geology and evolution is adopted (i.e., the hypothesis that no other principles of explanation should be used than can be observed as operative in the present)—then it is the statistical or mass model which is backed by empirical evidence. The business of the motivation and opinion researcher, statistical psychologist, etc., is based upon the premise that statistical laws obtain in human behavior; and that, for this reason, a small but well-chosen sample allows for extrapolation to the total population under consideration. The generally good working of a Gallup poll and prediction verifies the premise—with some incidental failure like the well-known example of the Truman election thrown in, as is to be expected with statistical predictions. The opposite contention—that history is governed by "free will" in the philosophical sense (i.e., rational decision for the better, the higher moral value or even enlightened self-interest) is hardly supported by fact. That here and there the statistical law is broken by "rugged individualists" is in its character. Nor does the role played in history by "great men" contradict the system concept in history; they can be conceived as acting like "leading parts," "triggers" or "catalyzers" in the historical process —a phenomenon well accounted for in the general theory of systems.

(5) A further question is the "organismic analogy" unanimously condemned by historians. They combat untiringly the "metaphysical," "poetical," "mythical" and thoroughly unscientific nature of Spengler's assertion that civilizations are a sort of

"organisms," being born, developing according to their internal laws and eventually dying. Toynbee (e.g., 1961) takes great pains to emphasize that he did not fall into Spengler's trap—even though it is somewhat difficult to see that his civilizations, connected by the biological relations of "affiliation" and "apparentation," even with a rather strict time span of development, are not conceived organismically.

Nobody should know better than the biologist that civilizations are not "organisms." It is trivial to the extreme that a biological organism, a material entity and unity in space and time, is something different from a social group consisting of distinct individuals, and even more from a civilization consisting of generations of human beings, of material products, institutions, ideas, values, and what not. It implies a serious underestimate of Vico's, Spengler's (or any normal individual's) intelligence to suppose that they did not realize the obvious.

Nevertheless, it is interesting to note that, in contrast to the historians' scruples, sociologists do not abhor the "organismic analogy" but rather take it for granted. For example, in the words of Rapoport and Horvath (1959):

There is some sense in considering a real organization as an organism, that is, there is reason to believe that this comparison need not be a sterile metaphorical analogy, such as was common in scholastic speculation about the body politic. Quasi-biological functions are demonstrable in organizations. They maintain themselves; they sometimes reproduce or metastasize; they respond to stresses; they age, and they die. Organizations have discernible anatomies and those at least which transform material inputs (like industries) have physiologies.

Or Sir Geoffrey Vickers (1957):

Institutions grow, repair themselves, reproduce themselves, decay, dissolve. In their external relations they show many characteristics of organic life. Some think that in their internal relations also human institutions are destined to become increasingly organic, that human cooperation will approach ever more closely to the integration of cells in a body. I find this prospect unconvincing (and) unpleasant. (N.B., so does the present author.)

And Haire (1959, p. 272):

The biological model for social organizations—and here, particularly for industrial organizations—means taking as a model the living organism and the processes and principles that regulate its growth and development. It means looking for lawful processes in organizational growth.

The fact that simple growth laws apply to social entities such as manufacturing companies, to urbanization, division of labor, etc., proves that in these respects the "organismic analogy" is correct. In spite of the historians' protests, the application of theoretical models, in particular, the model of dynamic, open and adaptive systems (McClelland, 1958) to the historical process certainly makes sense. This does not imply "biologism," i.e., reduction of social to biological concepts, but indicates system principles applying in both fields.

(6) Taking all objections for granted—poor method, errors in fact, the enormous complexity of the historical process—we have nevertheless reluctantly to admit that the cyclic models of history pass the most important test of scientific theory. The predictions made by Spengler in *The Decline of the West*, by Toynbee when forecasting a time of trouble and contending states, by Ortega y Gasset in *Revolt of the Masses*—we may as well add *Brave New World* and *1984*—have been verified to a disquieting extent and considerably better than many respectable models of the social scientists.

Does this imply "historic inevitability" and inexorable dissolution? Again, the simple answer was missed by moralizing and philosophizing historians. By extrapolation from the life cycles of previous civilizations nobody could have predicted the Industrial Revolution, the Population Explosion, the development of atomic energy, the emergence of underdeveloped nations, and the expansion of Western civilization over the whole globe. Does this refute the alleged model and "law" of history? No, it only says that this model—as every one in science—mirrors only certain aspects or facets of reality. Every model becomes dangerous only when it commits the "Nothing-but" fallacy which mars not only theoretical history, but the models of the mechanistic world picture, of psychoanalysis and many others as well.

We have hoped to show in this survey that General System Theory has contributed toward the expansion of scientific theory; has led to new insights and principles; and has opened up new problems that are "researchable," i.e., are amenable to further study, experimental or mathematical. The limitations of the theory and its applications in their present status are obvious; but the principles appear to be essentially sound as shown by their application in different fields.

5 The Organism Considered as Physical System

The Organism as Open System

Physical chemistry presents the theory of kinetics and equilibria in chemical systems. As example, consider the reversible reaction in ester formation:

$$C_2H_5OH + CH_3 \cdot COOH \rightleftarrows CH_3COO \cdot C_2H_5 + H_2O,$$

in which always a certain quantitative ratio between alcohol and acetic acid on the one hand, and between ester and water on the other, is established.

Application of physico-chemical equilibrium principles, especially of chemical kinetics and the law of mass action, has proved to be of fundamental importance for the explanation of physiological processes. An example is the function of blood, to transport oxygen from the lung to the tissues of the body and, conversely, carbon dioxide formed in the tissues to the lungs for exhalation; the process results from the equilibria between hemoglobin, oxyhemoglobin and oxygen according to the law of mass action, and quantitative formulations can be stated not only for the simple conditions in hemoglobin solution, but also for the more complicated ones in the blood of vertebrates. The importance of kinetic consideration of enzyme reactions, of respiration, fermentation, etc., is well known. Similarly, other physico-chemical equilibria (distribution, diffusion, adsorption, electrostatic equilibria) are of fundamental physiological significance (cf. Moser and Moser-Egg, 1934).

Considering the organism as a whole, it shows characteristics similar to those of systems in equilibrium (cf. Zwaardemaker, 1906, 1927). We find, in the cell and in the multicellular organism, a certain composition, a constant ratio of the components, which at first resembles the distribution of components in a chemical system in equilibrium and which, to a large extent, is maintained under different conditions, after disturbances, at different body size, etc.: an independence of composition of the absolute quantity of components, regulative capacity after disturbances, constancy of composition under changing conditions and with changing nutrition, etc. (cf. von Bertalanffy, 1932, pp. 190ff.; 1937, pp. 80ff.).

We realize at once, however, that there may be systems in equilibrium in the organism, but that the organism as such cannot be considered as an equilibrium system.

The organism is not a closed, but an open system. We term a system "closed" if no material enters or leaves it; it is called "open" if there is import and export of material.

There is, therefore, a fundamental contrast between chemical equilibria and the metabolizing organisms. The organism is not a static system closed to the outside and always containing the identical components; it is an open system in a (quasi-)steady state, maintained constant in its mass relations in a continuous change of component material and energies, in which material continually enters from, and leaves into, the outside environment.

The character of the organism as a system in steady (or rather quasi-steady) state is one of its primary criteria. In a general way, the fundamental phenomena of life can be considered as consequences of this fact. Considering the organism over a shorter span of time, it appears as a configuration maintained in a steady state by the exchange of components. This corresponds to the first main field of general physiology—i.e., physiology of metabolism in its chemical and energetic aspects. Superimposed on the steady state are smaller process waves, basically of two kinds. First there are periodic processes originating in the system itself and hence autonomic (e.g., automatic movements of the organs of respiration, circulation and digestion; automatic-rhythmic, electrical activities of nerve centers and the brain supposedly resulting from rhythmic chemical discharges; automatic movements of the organism as a whole). Secondly, the organism reacts to temporary changes in environment, to "stimuli," with reversible fluctuations of its steady state. This is the group of processes caused by changes of external conditions and hence heteronomic subsumed in physiology of excitation. They can be considered as temporary disturbances of the steady state from which the organism returns to "equilibrium," to the equal flow of the steady state. Such consideration has proved to be useful and leading to quantitative formulations (cf. p. 137). Finally, the definition of the state of the organism as steady state is valid only in first approximation, insofar as we envisage shorter periods of time in an "adult" organism, as we do, for example, in investigating metabolism. If we take the total life cycle, the process is not stationary but only quasi-stationary, subject to changes slow enough to abstract from them for certain research purposes, and comprising embryonic development, growth, aging, death, etc. These phenomena, not quite exhaustively encompassed under the term of morphogenesis, represent the third large complex of problems in general physiology. Such consideration proves especially useful in areas accessible to quantitative formulation.

In general, physical chemistry is limited almost exclusively to consideration of processes in closed systems. To these refer the well-known formulations of physical chemistry; the law of mass action, in particular, is used only for definition of true chemical equilibria in closed systems. The applicability of chemical equilibria to, e.g., transfer reactions is based on the fact that these are fast ionic reactions attaining equilibrium. Open chemical systems are hardly taken into consideration in physical chemistry. This restriction of kinetics to closed systems is understandable; for open systems are more difficult to establish technically, and not of major importance in the purely physical consideration. Nevertheless, such arrangements are easily visualizable—e.g., when in a reaction $a \rightleftarrows b$ the product b of the left-to-right reaction is continually removed from the system by suitable means (precipitation, dialysis through a membrane permeable only for b but not for a, etc.) while a is continually introduced into the system. Systems of this kind occasionally occur in technological chemistry; continuous fermentation in the production of acetic acid is an example for what is here called "open chemical system."

However, such systems are of great importance to the biologist.

For open chemical systems are indeed realized in nature in the form of living organisms, maintaining themselves in a continuous exchange of their components. "Life is a dynamic equilibrium in a polyphasic system" (Hopkins).

We therefore need a definition of the so-called stationary equilibrium, the constancy of composition in the change of components, similarly as well-known expressions of physical chemistry define true chemical equilibria in closed systems.

Obviously, the reaction system and reaction conditions are infinitely more complicated in organisms than in the systems usually dealt with in physical chemistry. These are reactions among an extraordinarily high number of components. Moreover, the cell and organism are not homogeneous systems (a true solution), but represent highly heterogeneous, colloidal systems so that reactions depend not only on mass action but on many physico-chemical factors of adsorption, diffusion, etc. Even enzyme reactions in the test tube do not, in general, simply follow the law of mass action. This being the case, it is clear that reactions even in simple organismic systems cannot be written in a closed system of equations; this is possible only for isolated partial systems. It is, however, possible, *first*, to state certain general principles for open systems, irrespective of the special nature of the system. *Secondly*, although in view of the enormous number of reactions in the organism and even the individual cell, it is impossible to follow individual reactions, expressions can be used that represent statistical averages of a multitude of incalculable or even unknown processes. Such a procedure is already applied in chemistry by using overall formulas for reactions consisting of numerous steps. Similarly, balance equations in physiology of metabolism and bioenergetics are based on statistical averages resulting from numerous and largely unknown processes in intermediary metabolism. We may, for instance, summarize anabolic and catabolic processes as "assimilation" and "dissimilation," respectively, and consider, as a first approximation, the steady state as balance of "assimilation" and "dissimilation." Such magnitudes, representing statistical averages of a multitude of inextricable processes, can be used for calculation in a way similar to that conventionally used in physical chemistry for individual compounds and reactions.

The maintenance of the system in a continuous flow and exchange of material and energy, the order of innumerable physico-chemical reactions in a cell or organism in a way granting the first, the maintenance of a constant ratio of the components even under different conditions, after disturbances, at different sizes, etc., are the central problems of organic metabolism. The double-faced change of living systems in assimilation and dissimilation manifests—in the words of von Tschermaks (1916)—a trend toward maintenance of a certain state, regeneration compensating the disturbance caused by degeneration. How is it that what has been lost in the process is rebuilt from the materials offered in nutrition, that building blocks liberated by enzymes find the right place in the organismic system so that it maintains itself in metabolism? What is the principle of "automatic self-regulation" of metabolism? We are possessed of a vast knowledge of physico-chemical processes in the cell and in the organism; but we must not overlook the fact "that even after complete explanation of individual processes, we are worlds away from fully understanding the total metabolism of a cell" (M. Hartmann, 1927, p. 258). Extremely little is known about the principles controlling the individual processes in the way indicated above. No wonder that again and again the problem led to vitalistic conclusions (e.g. Kottje, 1927).

Obviously, general principles as those we are going to develop cannot provide a detailed explanation of those problems; they can, however, indicate the general physical foundations of that essential characteristic of life, self-regulation of metabolism and maintenance in change of components. The special way in which these are realized in individual metabolic processes can be determined only by experimental investigation. It can be hoped, however, that the general consideration alerts to possibilities hitherto hardly envisaged, and that the formulations proposed, or similar equations, be apt to describe concrete individual phenomena.

General Characteristics of Open Chemical Systems

True equilibria in closed systems and stationary "equilibria" in open systems show a certain similarity, inasmuch as the system, taken as a whole and in view of its components, remains constant in both systems. But the physical situation in both cases is fundamentally different. Chemical equilibria in closed systems are based on reversible reactions; they are a consequence of the second principle of thermodynamics and are defined by minimum free energy. In open systems, in contrast, the steady state is not reversible as a whole nor in many individual reactions. Furthermore, the second principle applies, by definition, to closed systems only and does not define the steady state.

A closed system *must*, according to the second principle, eventually attain a time-independent state of equilibrium, defined by maximum entropy and minimum free energy (heat equilibrium, thermodynamic derivation of the law of mass action by van't Hoff, etc.), where the ratio between the phases remains constant. An open chemical system *may* attain (certain conditions presupposed) a time-independent steady state, where the system remains constant as a whole and in its (macroscopic) phases, though there is a continuous flow of component materials.

A closed system in equilibrium does not need energy for its preservation, nor can energy be obtained from it. For example, a closed reservoir contains a large amount of (potential) energy; but it cannot drive a motor. The same is true of a chemical system in equilibrium. It is not a state of chemical rest; rather reactions are continually going on, so regulated by the law of mass action that as much is formed of every species of molecules or ions as disappears. Nevertheless, the chemical equilibrium is incapable of performing work. For maintaining the processes going on, no work is required nor can work be won from it. The algebraic sum of work obtained from and used by the elementary reactions equals zero. In order to perform work, it is necessary that the system be not in a state of equilibrium but tend to attain it; only then can energy be won. In order that this is achieved continually, the hydrodynamic as well as chemical system must be arranged as stationary—i.e., a steady flow of water or chemical substances must be maintained whose energy content is transformed into work. Continuous working capacity is, therefore, not possible in a closed system which tends to attain equilibrium as soon as possible, but only in an open system. The apparent "equilibrium" found in an organism is not a true equilibrium incapable of performing work; rather it is a dynamic pseudo-equilibrium, kept constant at a certain distance from true equilibrium; so being capable of performing work but, on the other hand, requiring continuous import of energy for maintaining the distance from true equilibrium.

For the maintenance of "dynamic equilibrium," it is necessary that the rates of processes be exactly harmonized. Only in this way is it possible that certain components can be broken down, so liberating usable energy while, on the other hand, import prevents the system from attaining equilibrium. Fast reactions, also in the organism, lead to chemical equilibrium (e.g. of hemoglobin and oxygen); slow reactions do not reach equilibrium but are kept in a steady state. Therefore, the condition for the existence of a chemical system in a steady state is a certain slowness of reactions. Momentary reactions, like those between ions, lead to equilibrium in "infinitely short" time. The maintenance of a steady state in the organism is due to the fact that it is composed of complex carbon compounds; these are, on the one hand, rich in energy but chemically inert, so that the maintenance of considerable chemical potential is possible; on the other hand, rapid and regulated release of this amount of energy is performed by enzyme actions, so that a steady state is maintained.

For deriving conditions and characteristics of steady states we may use a general transport equation. Let Q_i be a measure of the i-th element of the system, e.g. a concentration or energy in a system of simultaneous equations. Its variation may be expressed by:

$$\frac{\partial Q_i}{\partial t} = T_i + P_i \qquad (5.1)$$

T_i represents the velocity of transport of the element Q_i in a volume element at a certain point of space, while P_i is the rate of production.

Many equations appearing in physics, biology and even sociology, can be considered as special cases of (5.1). For example, in molecular magnitude, the P_i are functions indicating the rate of reactions by which the substances Q_i are formed and destroyed; the T_i will have different forms depending on the system concerned. If, for example, no outer forces influence the masses, the T_i will be expressed by Fick's diffusion equation. In case T_i disappears, we have the usual equations for a set of reactions in a closed system; if P_i disappears, we have the simple diffusion equation where T_i has the form: $T_i = D_i \nabla^2 Q_i$, the Laplacian symbol ∇^2 representing the sum of the second partial derivatives in the space coordinates x, y, z, the D_i the diffusion coefficients. In biology, equations of this type are found, e.g., in growth; and they appear in sociology and population dynamics. In general, the rate of change of a population equals the population movement (immigration minus emigration) plus rate of reproduction (birth minus death rate).

In general, we therefore have a set of simultaneous partial differential equations. P_i as well as T_i will, in general, be nonlinear functions of Q_i and other system variables Q_j and furthermore functions of the space coordinates x, y, z and time t. For solving the equation, we must know the special form of the equations, and the initial and limiting conditions.

For our purpose, two considerations are important, which we may call temporal cross and longitudinal sections. The first problem is the maintenance in a steady state which, biologically, is the fundamental problem of metabolism. The second concerns changes of the system with respect to time, biologically expressed, e.g. as growth. Briefly we shall also mention a third problem— i.e., periodic changes as, in the organismic realm, are characteristic of autonomic processes such as automatic-rhythmical movements, etc. These three aspects correspond to the general problems of the three main fields of physiology (cf. pp. 121f.).

The problem of "longitudinal temporal section," of the changes of the system in time, will be answered by solution of differential equation of type (**5.1**).

As a simple example, consider an open chemical system, consisting of only one component Q, reaction material being continually imported and resulting reaction products removed. Let E be the amount of imported reaction material per time unit; k, the reaction constant according to the law of mass action; kQ, therefore, the turnover; then, presupposed the amount imported at the beginning is greater than that transformed, the concentration of the system will increase according to the equation:

$$\frac{dQ}{dt} = E - kQ. \tag{5.2}$$

As is easily seen, this is a special case of the general equation (**5.1**). Since inflow was assumed to be constant and outflow equal to the chemical reaction, hence diffusion and concentration gradients were neglected (or, as may be said, a complete "stirring" of the system was assumed), the space co-ordinates in (5.1) disappear; instead of a partial, we have an ordinary differential equation. Concentration at time t then is:

$$Q = \frac{E}{k} - \left(\frac{E}{k} - Q_0\right) e^{-kt}, \qquad (5.3)$$

Q_0 being the initial concentration at $t = 0$. Concentration therefore asymptotically increases to a certain limit where turnover equals inflow (assumed to be constant). This maximal concentration is $Q_\infty = E/k$.

A system more approaching biological conditions is as follows. Let there be transport of material a_1 into the system proportional to the difference between its concentration outside and inside of the system $(X-x_1)$. Biologically, we may here think of simple sugars or amino acids. The imported material a_1 may form, in a monomolecular and reversible reaction, a compound a_2 of concentration x_2 (e.g. monosaccharids transformed into polysaccharids, amino acids into proteins). On the other hand, the substance a_1 may be catabolized in an irreversible reaction (e.g. oxidation, desamination) into a_3; and a_3 may be removed from the system, proportional to its concentration. Then we have the following system of reactions:

$$X \xrightarrow{K_1} x_1 \underset{k_2}{\overset{k_1}{\rightleftarrows}} x_2$$

$$k_3 \Big\downarrow$$

$$x_3$$

$$K_2 \downarrow$$

Outflow

and equations:

$$\frac{dx_1}{dt} = K_1(X - x_1) - k_1x_1 + k_2x_2 - k_3x_1 = x_1$$
$$(- K_1 - k_1 - k_3) + k_2x_2 + K_1X$$

$$\frac{dx_2}{dt} = k_1x_1 - k_2x_2 \qquad\qquad (5.4)$$

$$\frac{dx_3}{dt} = k_3x_1 - K_2x_3.$$

For eliminating the constant in the first equation, equate it to 0; x_1^*, x_2^* ... be the roots of these equations. We introduce as new variables:

$$x'_1 = x_1^* - x_1. \ldots \tag{5.5}$$

and reformulate (5.4) accordingly.

The general type of such equations is:

$$\left.\begin{aligned}
\frac{dx'_1}{dt} &= a_{11}x'_1 + a_{12}x'_2 + \ldots + a_{1n}x'_n \\
\frac{dx'_2}{dt} &= a_{21}x'_1 + a_{22}x'_2 + \ldots + a_{2n}x'_n \\
&\cdots\cdots\cdots\cdots\cdots\cdots\cdots\cdots\cdots\cdots \\
\frac{dx'_n}{dt} &= a_{n1}x'_1 + a_{n2}x'_2 + \ldots + a_{nn}x'_n
\end{aligned}\right\} \tag{5.6}$$

with the general solution (cf. p. 58):

$$\left.\begin{aligned}
x'_1 &= C_{11}e^{\lambda_1 t} + C_{12}e^{\lambda_2 t} + \ldots C_{1n}e^{\lambda_n t} \\
x'_2 &= C_{21}e^{\lambda_1 t} + C_{22}e^{\lambda_2 t} + \ldots C_{2n}e^{\lambda_n t} \\
&\cdots\cdots\cdots\cdots\cdots\cdots\cdots\cdots\cdots\cdots \\
x'_n &= C_{n1}e^{\lambda_1 t} + C_{n2}e^{\lambda_2 t} + \ldots C_{nn}e^{\lambda_n t}
\end{aligned}\right\} \tag{5.7}$$

The λ's are given by the characteristic equation:

$$\begin{vmatrix}
a_{11} - \lambda & a_{12} & \ldots & a_{1n} \\
a_{21} & a_{22} - \lambda & \ldots & a_{2n} \\
\multicolumn{4}{c}{\cdots\cdots\cdots\cdots\cdots\cdots\cdots\cdots} \\
a_{n1} & a_{n2} & \ldots & a_{nn} - \lambda
\end{vmatrix} = 0. \tag{5.8}$$

We now consider the temporal cross-section, i.e. the distribution of components in the time-independent steady state.

In general, a system defined by equation (5.1) can have three different solutions. First, there can be unlimited increase of the Q_i; secondly, a time-independent steady state may be attained; thirdly, there may be periodic solutions.

It is difficult to prove the existence of a steady state for the general system (5.1), yet it can be shown in certain cases. Suppose that both terms are linear in the Q_i and independent of t. Then the solution can be found by standard methods of integration and is of the form:

$$Q_i = Q_{i1}(x, y, z) + Q_{i2}(x, y, z, t), \tag{5.9}$$

where Q_{i2} is a function of t which with increasing time is decreasing to zero for certain relations between constants and limiting conditions.

If, on the other hand, there is a time-independent steady state expressed by Q_{i1} in(5.9), Q_{i1} must suffice the time-independent equation:

$$T_i + P_i = 0 \qquad (5.10)$$

From this we see:

(1) If there is a stationary solution, the composition of the system in the steady state remains constant with respect to the components Q_i although the reactions continue and do not reach equilibrium as in a closed system, and although there is inflow and outflow of material; the situation so highly characteristic of organismic systems.

(2) In the steady state the number of elements entering state Q_i (x, y, z, t) by transport and chemical reaction per time unit equals the number leaving it.

Similar considerations can be made with respect to periodic solutions. It is true that the above derivation presupposes rather special assumptions on the nature of the equations. However, although no general criterion is known for the existence of stationary and periodic solutions in system (5.1), these conditions can be indicated for certain types of linear and even non-linear cases. Important to us is the fact that the existence of stationary, dynamic "equilibria" in open systems, or as we may also say, the existence of a certain order of processes guaranteed by dynamic rather than structural-mechanical principles, can be derived from general considerations.

Solving equations (5.4) for the steady state we obtain:

$$x_1 : x_2 : x_3 = 1 : \frac{k_1}{k_2} : \frac{k_3}{K_2}.$$

We therefore see that in the steady state a constant ratio between the components is established although it is not, as in a closed system, based on an equilibrium of reversible reactions, but the reactions are partly irreversible. Moreover, the ratio of components in the steady state depends only on the reaction constants, not on the amount of the inflow; the system thus shows "self-regulation," comparable to organismic systems, where the ratio of the components is maintained in changing inflow, changing absolute size, etc.

Furthermore we find:

$$x_1{}^* = \frac{K_1 X}{K_1 + k_3}.$$

In case an external disturbance ("stimulus") leads to increased catabolism—e.g., increase of the reaction constant k_3 while the other constants remain unaltered—x_1 decreases. Since, however, inflow is proportional to the concentration difference $X - x_1$, with increase of the latter intake is increased. If, after cessation of the "stimulus," the constant of catabolism returns to its normal value, the system will return to its original state. If, however, the disturbance and hence the change of rate of catabolism persists, a new steady state will be established. Thus the system develops forces directed against the disturbance, tending to compensate increased catabolism by increased intake. It therefore shows "adaptation" to the new situation. These, too, are "self-regulative" characteristics of the system.

It can, therefore, be seen that the properties indicated as characteristic of organismic systems, are consequences of the nature of open systems: maintenance in "dynamic equilibrium," independence of composition of the absolute quantity of components, maintenance of the composition under changing conditions and nutrition, reestablishment of dynamic equilibrium after normal catabolism or catabolism increased by a stimulus, dynamic order of processes, etc. "Self-regulation of metabolism" can be made understandable on the basis of physical principles.

Equifinality

One important characteristic of biological systems is circumscribed by terms like "purposiveness," "finality," "goal-seeking," etc. Let us see whether physical considerations can contribute to a clarification of these terms.

It has often been emphasized that every system attaining an equilibrium shows, in a certain way, "finalistic" behavior as was discussed previously (pp. 75f.).

More important is the following consideration. Frequent attempts have been made to understand organic regulations as establishment of an "equilibrium" (of course, of extremely complicated nature) (e.g., Köhler, 1927), to apply LeChatelier's and similar principles. We are not in a position to define such "equilibrium state" in complicated organic processes, but we can easily see that such a conception is, in principle, inadequate. For, apart from certain individual processes, living systems are not closed systems in true equilibrium but open systems in a steady state.

Nevertheless, steady states in open systems have remarkable characteristics.

An aspect very characteristic of the dynamic order in organismic processes can be termed as *equifinality*. Processes occurring in machine-like structures follow a fixed pathway. Therefore the final state will be changed if the initial conditions or the course of processes is altered. In contrast, the same final state, the same "goal," may be reached from different initial conditions and in different pathways in organismic processes. Examples are the development of a normal organism from a whole, a divided or two fused ova, or from any pieces as in hydroids or planarians, or the reaching of a definite final size from different initial sizes and after a different course of growth, etc.

We may define:

A system of elements Q_i (x, y, z, t) is equifinal in any subsystem of elements Q_j, if the initial conditions Q_{io} (x, y, z) can be changed without changing the value of Q_j (x, y, z, ∞).

We can stipulate two interesting theorems:

1. If there exists a solution of form (**5.9**), initial conditions do not enter into the solution for the steady state. This means: *If open systems* (of the kind discussed) attain a steady state, this has a *value equifinal or independent of initial conditions*. A general proof is difficult because of the lack of general criteria for the existence of steady states; but it can be given for special cases.

2. In a closed system, some function of the elements—e.g., total mass or energy—is by definition a constant. Consider such an integral of the system, $M(Q_i)$. If the initial conditions of Q_i are given as Q_{io}, we must have:

$$M(Q_i) = M(Q_{io}) = M, \qquad (5.11)$$

independent of t. If the Q_i, tend toward an asymptotic value, Q_{i1},

$$M(Q_{i1}) = M \qquad (5.12)$$

M, however, cannot be entirely independent of Q_{io}; with change of Q_{io}, also M and therefore $M(Q_{i1})$ are altered. If this integral changes its value, at least some of the Q_{i1} must also change. This, however, is contrary to the definition of equifinality. We may therefore stipulate the theorem: *A closed system cannot be equifinal with regard to all Q_i*.

For example, in the simplest case of an open chemical system according to equation (5.2), concentration at time t is given by (5.3); for $t = \infty$, $Q = E/k$, i.e. it is independent of the initial concentration Q_o and dependent only on the system constants E and k. A derivation of equifinality—i.e., the reaching of a steady state independent of time and initial conditions—in diffusion systems can be found in Rashevsky (1938, Chapter 1).

The general consideration, of course, does not provide an explanation for specific phenomena if we do not know the special conditions. Yet, the general formulation is not without interest. We see, first, that it is possible to give a physical formulation to the apparently metaphysical or vitalistic concept of finality; as is well known, the phenomenon of equifinality is the basis of the so-called "proofs" of vitalism of Driesch. Secondly, we see the close relation between one fundamental characteristic of the organism, i.e. the fact that it is not a closed system in thermodynamic equilibrium but an open system in a (quasi-)stationary state with another one, equifinality.[1]

A problem not here considered is the dependence of a system not only on actual conditions, but also on past conditions and the course taken in the past. These are the phenomena known as "after-effect," "hereditary" (in mathematical sense: E. Picard) or "historic" (Volterra) (cf. D'Ancona, 1939, Chapter XXII). In this category belong phenomena of hysteresis in elasticity, elec-

---

[1]The limitations of organismic regulation are based on the fact that the organism (ontogenetically as well as phylogenetically) passes from the state of a *system* of dynamically interacting elements to the state of structural "mechanisms" and individual causal chains (cf. pp. 68ff.). If the components become independent of each other, the change in each one depends only on the conditions within this component. Change or removal of a component must cause a final state different from the normal state; regulation is impossible in a completely "mechanized" system disintegrated into mutually independent causal chains (except for control by feedback mechanisms cf. p. 42ff. and elsewhere).

tricity, magnetism, etc. Taking dependence on the past into consideration, our equations would become integro-differential equations as discussed by Volterra (cf. D'Ancona) and Donnan (1937).

Biological Applications

It should have become evident by now that many characteristics of organismic systems, often considered vitalistic or mystical, can be derived from the system concept and the characteristics of certain, rather general system equations, in connection with thermodynamic and statistical-mechanical considerations.

If the organism is an open system, the principles generally applying to systems of this kind *must* apply to it (maintenance in change, dynamic order of processes, equifinality, etc.) quite irrespective of the nature of the obviously extremely complicated relations and processes between the components.

Naturally, such a general consideration does not give an explanation for particular life phenomena. The principles discussed should, however, provide a general frame or scheme within which quantitative theories of specific life phenomena should be possible. In other terms, theories of individual biological phenomena should turn out as special cases of our general equations. Without striving for completeness, a few examples may show that and how the conception of organism as open chemical system and steady state has proved an efficient working hypothesis in various fields.

Rashevsky (1938) investigated, as a highly simplified theoretical model of a cell, the behavior of a metabolizing droplet into which substances diffuse from outside, in which they undergo chemical reactions, and from which reaction products flow out. This consideration of a simple case of open system (whose equations are special cases of our equation [**5.1**]) allows mathematical deduction of a number of characteristics always considered as essential life phenomena. There results an order of magnitude for such systems corresponding to that of actual cells, growth and periodic division, the impossibility of spontaneous generation (*omnis cellula e cellula*), general characteristics of cell division, etc.

Osterhout (1932–33) applied, and quantitatively elaborated, the open-system consideration to phenomena of permeability. He studied permeation in cell models consisting of a non-aqueous layer surrounded by an aqueous outer and inner fluid (the latter corresponding to cell sap). An accumulation of penetrating substances takes place within this cell, explained by salt formation of the penetrating substance. The result is not an equilibrium but a steady state, in which the composition of the cell sap remains constant under increase of volume. This model is similar to that mentioned on p. 126. Mathematical expressions were derived, and the kinetics of this model is similar to that in living cells.

Open systems and steady states generally play a fundamental role in metabolism although mathematical formulation has been possible only in simple cases or models. For example, the continuation of digestion is only possible because of the continuous resorption of the products of enzymatic action by the intestine; it therefore never reaches a state of equilibrium. In other cases, accumulation of reaction products may lead to stopping the reaction which explains some regulatory processes (cf. von Bertalanffy, 1932, p. 191). This is true of the use of depot materials: Decomposition of starch stored in the endosperm of many plant seeds into soluble products is regulated by the need of the growing plant for carbohydrates; if development is experimentally inhibited, the use of starch in the endosperm stops. Pfeffer and Hansteen (quoted from Höber, 1926, p. 870) made it probable that the accumulation of sugar originating from digestion of starch and not used up by the inhibited seedling is the cause for the stopping of starch breakdown in the endosperm. If the endosperm is isolated and connected with a small plaster column, the breakdown of starch continues in the endosperm if the sugar diffuses through the plaster column into a quantity of water, but is inhibited if the column is placed in a small quantity of water only so that the concentration of sugar inhibits hydrolysis.

One field where processes can already be formulated in the form of equations, is the theory of growth. It can be assumed (von Bertalanffy, 1934), that growth is based on a counteraction of anabolic and catabolic processes: The organism grows when building-up surpasses breaking-down, and becomes stationary, when both processes are balanced. It can further be assumed that, in many organisms, catabolism is proportional to volume (weight), anabolism is proportional to resorption, i.e., a surface.

This hypothesis can be supported by a number of morphological and physiological arguments and in simple cases, such as planarians, can be partly verified by measurement of intestinal surface (von Bertalanffy, 1940b). If κ is a constant for catabolism per unit mass, total catabolism will be κw (w = weight); similar, with η as constant per unit surface, anabolism will be ηs, and weight increase defined by the difference of these magnitudes:

$$\frac{dw}{dt} = \eta s - \kappa w. \tag{5.13}$$

From this basic equation, expressions can be derived which quantitatively represent empirical growth curves and explain a considerable number of growth phenomena. In simpler cases these growth laws are realized with the exactness of physical experiments. Moreover, the rate of catabolism can be calculated from growth curves and comparing values so calculated with those directly determined in physiological experiment, an excellent agreement is found. This tends to show, first, that the parameters of the equations are not mathematically constructed entities but physiological realities; secondly, that basic processes of growth are rendered by the theory (cf. Chapter 7).

This example well illustrates the principle of equifinality discussed previously. From (5.13) follows for weight increase:

$$w = \left[\frac{E}{k} - \left(\frac{E}{k} - \sqrt[3]{w_0} \right) e^{-kt} \right]^3, \tag{5.14}$$

where E and k are constants related to η and κ, and where w_0 is the initial weight. The stationary final weight is given by $w^* = (E/k)^3$; it is thus independent of the initial weight. This can also be shown experimentally since the same final weight, defined by the species-specific constants E and k, may be reached after a growth curve entirely different from the normal one (cf. von Bertalanffy, 1934).

Obviously, this growth theory follows the conceptions of kinetics of open systems; equation (5.13) is a special case of the general equation (5.1). The basic characteristic of the organism, its representing an open system, is claimed to be the principle of organismic growth.

Another field where this concept has proved itself fruitful is the phenomenon of excitation. Hering first considered the phenomena of irritability as reversible disturbances of the stationary flow of organismic processes. In the state of rest, assimilation and dissimilation are balanced; a stimulus causes increased dissimilation; but then the quantity of decomposable substances is decreased, the counteracting assimilation process is accelerated, until a new steady state between assimilation and dissimilation is reached. This theory has proved to be extremely fruitful. The theory of Pütter (1918–1920), further developed by Hecht (1931), considers the formation of excitatory substances from sensitive substances (e.g., visual purple in the rods of the vertebrate eye) and their disappearance as the basis of excitation. From the counteraction of these processes, production and removal of excitatory substances, the quantitative relations of sensory excitation can be derived on the basis of chemical kinetics and the law of mass action: threshold phenomena, adaptation to light and darkness, intensity discrimination, Weber's law and its limitations, etc. A similar hypothesis of excitatory and inhibitory substances and of a dissimilation mechanism under the influence of stimuli forms the basis of Rashevsky's theory (1933) of nervous excitation by electric stimuli, formally identical with the theory of excitation by Hill (1936). The theory of excitatory substances is not limited to sense organs and the peripheral nervous system, but applicable also to the transmission of excitation from one neuron to another at the synapses. Without entering the still unsettled question of a chemical or electrical theory of transmission in the central nervous system, the first explains many of the basic features of the central nervous system compared with the peripheral nerve, such as irreciprocity of conduction, retardation of transmission in the central nervous system, summation and inhibition; here, too, is the possibility of quantitative formulations. Lapicque, e.g., developed a mathematical theory of summation in the central nervous system; according to Umrath, it can be interpreted by the production and disappearance of excitatory substances.

We may therefore say, first, that the large areas of metabolism, growth, excitation, etc., begin to fuse into an integrated theoretical field, under the guidance of the concept of open systems; secondly, that a large number of problems and possible quantitative formulations result from this concept.

In connection with the phenomena of excitation, it should be mentioned that this conception also is significant in pharmacological problems. Loewe (1928) applied the concept of the organism as open system in quantitative analysis of pharmacological effects and derived the quantitative relations for the action mechanism of certain drugs ("put-in," "drop-in," "block-out" systems).

Finally, problems similar to those discussed with respect to the individual organism also occur with respect to supra-individual entities which, in the continual death and birth, immigration and emigration of individuals, represent open systems of a higher nature. As a matter of fact, the equations developed by Volterra for population dynamics, biocoenoses, etc. (cf. D'Ancona, 1939) belong to the general type discussed above.

In conclusion, it may be said that consideration of organismic phenomena under the conception discussed, a few general principles of which have been developed, has already proved its importance for explanation of specific phenomena of life.

6 The Model of Open System

The Living Machine and Its Limitations

The present discussion may be started with one of those trivial questions which are often only too difficult to answer scientifically. What is the difference between a normal, a sick and a dead organism? From the standpoint of physics and chemistry the answer is bound to be that the difference is not definable on the basis of so-called mechanistic theory. Speaking in terms of physics and chemistry, a living organism is an aggregate of a great number of processes which, sufficient work and knowledge presupposed, can be defined by means of chemical formulas, mathematical equations, and laws of nature. These processes, it is true, are different in a living, sick or dead dog; but the laws of physics do not tell a difference, they are not interested in whether dogs are alive or dead. This remains the same even if we take into consideration the latest results of molecular biology. One DNA molecule, protein, enzyme or hormonal process is as good as another; each is determined by physical and chemical laws, none is better, healthier or more normal than the other.

Nevertheless, there is a fundamental difference between a live and a dead organism; usually, we do not have any difficulty in distinguishing between a living organism and a dead object. In a living being innumerable chemical and physical processes are so "ordered" as to allow the living system to persist, to grow, to develop, to reproduce, etc. What, however, does this notion of "order" mean, for which we would look in vain in a textbook of physics? In order to define and explain it we need a model, a conceptual construct. One such model was used since the beginnings of modern science. This was the model of the living machine. Depending on the state of the art, the model found different interpretations. When, in the seventeenth century, Descartes introduced the concept of the animal as a machine, only *mechanical machines* existed. Hence the animal was a complicated clockwork. Borelli, Harvey and other so-called iatrophysicists explained the functions of muscles, of the heart, etc., by mechanical principles of levers, pumps and the like. One can still see this in the opera, when in the *Tales of Hoffmann* the beautiful Olympia turns out to be an artfully constructed doll, an automaton as it was called at the time. Later, the steam engine and thermodynamics were introduced, which led to the organism being conceived as a *heat engine*, a notion which lead to caloric calculations and other things. However, the organism is not a heat engine, transforming the energy of fuel into heat and then into mechanical energy. Rather it is a *chemodynamic machine*, directly transforming the energy of fuel into effective work, a fact on which, for example, the theory of muscle action is based. Lately, self-regulating machines came to the fore, such as thermostats, missiles aiming at a target and the servomechanisms of modern technology. So the organism became a *cybernetic machine*, explanatory of many homeostatic and related phenomena. The most recent development is in terms of *molecular machines*. When one talks about the "mill" of the Krebs cycle of oxidation or about the mitochondria as "power plant" of the cell, it means that machinelike structures at the molecular level determine the order of enzyme reactions; similarly, it is a micromachine which transforms or translates the genetic code of DNA of the chromosomes into specific proteins and eventually into a complex organism.

Notwithstanding its success, the machine model of the organism has its difficulties and limitations.

First, there is the problem of *the origin of the machine*. Old Descartes did not have a problem because his animal machine was the creation of a divine watchmaker. But how do machines come about in a universe of undirected physico-chemical events? Clocks, steam engines and transistors do not grow by themselves in nature. Where do the infinitely more complicated living machines come from? We know, of course, the Darwinistic explanation; but a doubt remains, particularly in the physically minded; there remain questions not usually posed or answered in textbooks on evolution.

Secondly, there is the problem of *regulation*. To be sure, self-repairing machines are conceivable in terms of the modern theory of automata. The problem comes in with regulation and repair after arbitrary disturbances. Can a machine, say, an embryo or a brain, be programmed for regulation not after a certain disturbance or finite set of disturbances, but after disturbances of an indefinite number? The so-called Turing machine can, in principle, resolve even the most complex process into steps which, if their number is finite, can be reproduced by an automaton. However, the number of steps may be neither finite nor infinite, but "immense," i.e., transcending the number of particles or possible events in the universe. Where does this leave the organism as machine or automaton? It is well-known that organic regulations of such sort were used by vitalists as proof that the organic machine is controlled and repaired by superphysical agents, so-called entelechies.

Even more important is a third question. The living organism is maintained in a continuous *exchange of components*; metabolism is a basic characteristic of living systems. We have, as it were, a machine composed of fuel spending itself continually and yet maintaining itself. Such machines do not exist in present-day technology. In other words: A machinelike structure of the organism cannot be the ultimate reason for the order of life processes because the machine itself is maintained in an ordered flow of processes. The primary order, therefore, must lie in the process itself.

Some Characteristics of Open Systems

We express this by saying that living systems are basically open systems (Burton, 1939; von Bertalanffy, 1940a; Chapter 5). An open system is defined as a system in exchange of matter with its environment, presenting import and export, building-up and breaking-down of its material components. Up to comparatively recent times physical chemistry, in kinetics and thermodynamics, was restricted to closed systems; the theory of open systems is relatively new and leaves many problems unsolved. The development of kinetic theory of open systems derives from two sources: first the biophysics of the living organism, secondly developments in industrial chemistry which, besides reactions in closed containers or batch processes, increasingly uses continuous reaction systems because of higher efficiency and other advantages. The thermodynamic theory of open systems is the so-called irreversible thermodynamics (Meixner & Reik, 1959); it became an important generalization of physical theory through the work of Meixner, Onsager, Prigogine and others.

Even simple open systems show remarkable characteristics (Chapter 5). Under certain conditions, open systems approach a time-independent state, the so-called steady state (*Fliessgleichgewicht* after von Bertalanffy, 1942). The steady state is maintained in distance from true equilibrium and therefore is capable of doing work; as it is the case in living systems, in contrast to systems in equilibrium. The system remains constant in its composition, in spite of continuous irreversible processes, import and export, building-up and breaking-down, taking place. The steady state shows remarkable regulatory characteristics which become evident particularly in its equifinality. If a steady state is reached in an open system, it is independent of the initial conditions, and determined only by the system parameters, i.e., rates of reaction and transport. This is called *equifinality* as found in many organismic processes, e.g., in growth (FIG. **6.1**). In contrast to closed

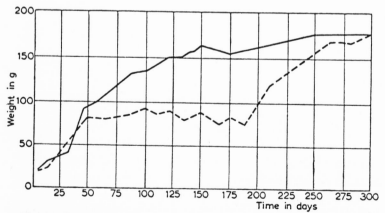

Fig. 6.1. Equifinality of growth. Heavy curve: normal growth of rats. Broken curve: at the 50th day, growth was stopped by vitamin deficiency. After reestablishment of normal regime, the animals reached the normal final weight. (After Höber from von Bertalanffy, 1960b).

physico-chemical systems, the same final state can therefore be reached equifinally from different initial conditions and after disturbances of the process. Furthermore, the state of chemical equilibrium is independent of catalyzers accelerating the processes. The steady state, in contrast, depends on catalyzers present and their reaction constants. In open systems, phenomena of *overshoot* and *false start* (FIG. **6.2**) may occur, with the system

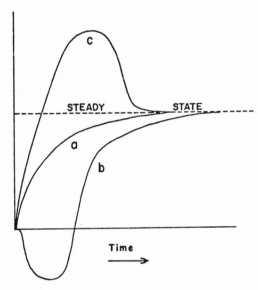

Fig. 6.2. Asymptotic approach to steady state (*a*), false start (*b*), and overshoot (*c*), in open systems. Schematic.

proceeding first in a direction opposite to that eventually leading to the steady state. Conversely, phenomena of overshoot and false start, as frequently found in physiology, may indicate that we are dealing with processes in open systems.

From the viewpoint of thermodynamics, open systems can maintain themselves in a state of high statistical improbability, of order and organization.

According to the second principle of thermodynamics, the general trend of physical processes is toward increasing entropy, i.e., states of increasing probability and decreasing order. Living systems maintain themselves in a state of high order and improbability, or may even evolve toward increasing differentiation and organization as is the case in organismic development and evolution. The reason is given in the expanded entropy function of Prigogine. In a closed system, entropy always increases according to the Clausius equation:

$$dS \geqq 0 \qquad (6.1)$$

In an open system, in contrast, the total change of entropy can be written according to Prigogine:

$$dS = d_e S + d_i S, \qquad (6.2)$$

$d_e S$ denoting the change of entropy by import, $d_i S$ the production of entropy due to irreversible processes in the system, such as chemical reactions, diffusion, heat transport, etc. The term $d_i S$ is always positive, according to the second principle; $d_e S$, entropy transport, may be positive or negative, the latter, e.g., by import of matter as potential carrier of free energy or "negative entropy." This is the basis of the negentropic trend in organismic systems and of Schrödinger's statement that "the organism feeds on negative entropy."

More complex open-system models, approximating biological problems, have been developed and analyzed by Burton, Rashevsky, Hearon, Reiner, Denbigh and other authors. In recent years, computerization has been widely applied for the solution of sets of numerous simultaneous equations (frequently nonlinear) (e.g., Franks, 1967; B. Hess and others) and for the *simulation* of complex open-system processes in physiological problems (e.g., Zerbst and coworkers; 1963 ff.). *Compartment theory* (Rescigno and Segre, 1967; Locker, 1966b) provides sophisticated methods for cases where reactions take place not in a homogenous space but in subsystems partly permeable to the reactants, as is the case in industrial systems and obviously many processes in the cell.

As can be seen, open systems compared with conventional closed systems show characteristics which seem to contradict the usual physical laws, and which were often considered as vitalistic characteristics of life, i.e., as a violation of physical laws, explainable only by introducing soul-like or entelechial factors into the organic happening. This is true of the equifinality of organic regulations, if, for example, the same "goal," a normal organism, is produced by a normal, a divided, two fused ova, etc. In fact, this was the most important "proof of vitalism" according to Driesch. Similarly, the apparent contradiction of the trend toward increase of entropy and disorder in physical nature, and the negentropic trend in development and evolution were often used as vitalistic arguments. The apparent contradictions disappear with the expansion and generalization of physical theory to open systems.

Open Systems in Biology

The model of open systems is applicable to many problems and fields of biology (Beier, 1962, 1965; Locker *et al.*, 1964, 1966a). A survey of the biophysics of open systems, including theoretical foundations and applications, was given some years ago (von Bertalanffy, 1953a). The present survey is restricted to some representative examples.

There is, first, the large field of Goethe's *Stirb und werde,* the continuous decay and regeneration, the dynamic structure of living systems at all levels of organization (Tables 6.1–6.3). Gen-

Table 6.1

Turnover rates of intermediates of cellular metabolism. (After B. Hess 1963)

structure	species	organ	turnover time in seconds
mitochondria	mouse	liver	1.3×10^6
hemoglobin	man	erythrocytes	1.5×10^7
aldolase	rabbit	muscle	1.7×10^6
pseudocholinesterase	man	serum	1.2×10^6
cholesterin	man	serum	9.5×10^5
fibrinogen	man	serum	4.8×10^4
glucose	rat	total organism	4.4×10^3
methionine	man	total organism	2.2×10^3
ATP glycolysis	man	erythrocytes	1.6×10^3
ATP glycolysis + respiration	man	thrombocytes	4.8×10^2
ATP glycolysis + respiration	mouse	ascites tumor	4.0×10^1
citrate cycle intermediates	rat	kidney	$1 \,-10$
glycolytic intermediates	mouse	ascites tumor	$0.1-\ 8.5$
flavoprotein$_{red.}$/flavoprotein$_{ox.}$	mouse	ascites tumor	4.6×10^{-2}
Fe^{2+}/Fe^{3+} – cytochrome a	grasshopper	wing muscle	10^{-2}
Fe^{2+}/Fe^{3+} – cytochrome a$_3$	mouse	ascites tumor	1.9×10^{-3}

Table 6.2

Protein turnover determined by introduction of glycine labelled with 15N.
(After SPRINSON & RITTENBERG (1949b)

		turnover rate (r)
RAT:		
	total protein	0.04
	proteins of liver, plasma and internal organs	0.12
	rest of body	0.033
MAN:		
	total protein	0.0087
	proteins of liver and serum	0.0693
	protein of musculature and other organs	0.0044

Table 6.3

Rate of mitosis in rat tissues. (After F. D. BERTALANFFY 1960)

	daily rate of mitosis (per cent)	renewal time (days)
Organs without mitosis		
nerve cells, neuroepithelium, neurilemma, retina, adrenal medulla	0	–
Organs with occasional mitosis but no cell renewal		
liver parenchyma, renal cortex and medulla, most glandular tissue, urethra, epididymis, vas deferens, muscle, vascular endothelium, cartilage, bone	less than 1	–
Organs with cell renewal		
upper digestive tract	7 –24	4.3–14.7
large intestine and anus	10 –23	4.3–10
stomach and pylorus	11 –54	1.9– 9.1
small intestine	64 –79	1.3– 1.6
trachea and bronchus	2 – 4	26.7–47.6
ureter and bladder	1.6– 3	33 –62.5
epidermis	3 – 5	19.1–34.5
sebaceous glands	13	8
cornea	14	6.9
lymph node	14	6.9
pulmonary alveolar cells	15	6.4
seminiferous epithelium	–	16

erally it may be said that this regeneration takes place at far higher turnover rates than was anticipated. For example, it is certainly surprising that calculation on the basis of open system revealed that the proteins of the human body have a turnover time of not much more than a hundred days. Essentially the same is true for cells and tissues. Many tissues of the adult organism are maintained in a steady state, cells being continuously lost by desquamation and replaced by mitosis (F. D. Bertalanffy and Lau, 1962). Techniques such as the application of colchicine that arrests mitosis and thus permits counting of dividing cells over certain periods, as well as labelling with tritiated thymidine, have revealed a sometimes surprisingly high renewal rate. Prior to such investigations, it was hardly expected that cells in the digestive tract or respiratory system have a life span of only a few days.

After the exploration of the paths of individual metabolic reactions, in biochemistry, it has now become an important task to understand integrated metabolic systems as functional units (Chance *et al.*, 1965). The way is through physical chemistry of enzyme reactions as applied in open systems. The complex network and interplay of scores of reactions was clarified in functions such as photosynthesis (Bradley and Calvin, 1956), respiration (B. Hess and Chance, 1959; B. Hess, 1963) and glycolysis, the latter investigated by a computer model of some hundred nonlinear differential equations (B. Hess, 1969). From a more general viewpoint, we begin to understand that besides visible morphologic organization, as observed by the electron microscope, light microscope and macroscopically, there is another, invisible, organization resulting from interplay of processes determined by rates of reaction and transport and defending itself against environmental disturbances.

Hydrodynamic (Burton, 1939; Garavaglia *et al.*, 1958; Rescigno, 1960) and particularly electronic analogs provide another approach besides physiological experiment, especially permitting solutions of multivariable problems which otherwise exceed time limits and available mathematical techniques. In this way Zerbst *et al.* (1963ff.) arrived at important results on temperature adaptation of heart frequency, action potentials of sensory cells (amending the Hodgkin-Huxley feedback theory), etc.

Furthermore, energetic conditions have to be taken into account. The concentration, say, of proteins in an organism does not correspond to chemical equilibrium; energy expense is necessary for the maintenance of the steady state. Thermodynamic consideration permits an estimate of energy expense and comparison with the energy balance of the organism (Schulz, 1950; von Bertalanffy, 1953a).

Another field of investigation is active transport in the cellular processes of import and export, kidney function, etc. This is connected with bioelectrical potentials. Treatment requires application of irreversible thermodynamics.

In the human organism, the prototype of open system is the blood with its various levels of concentrations maintained constant. Concentrations and removal of both metabolites and administered test substances follow open-systems kinetics. Valuable clinical tests have been developed on this basis (Dost, 1953–1962). In a broader context, pharmacodynamic action in general represents processes taking place when a drug is introduced into the open system of the living organism. The model of the open system can serve as foundation of the laws of pharmacodynamic effects and dose-effect relations (Loewe, 1928; Druckrey and Kuepfmüller, 1949; G. Werner, 1947; Dost, 1968).

Furthermore, the organism responds to external stimuli. This can be conceived of as disturbance and subsequent reestablishment of a steady state. Consequently, quantitative laws in sensory physiology, such as the Weber-Fechner law, belong to open systems kinetics. Hecht (1931), long before the formal introduction of open systems, expressed the theory of photoreceptors and existing laws in the form of "open" reaction kinetics of sensitive material.

The greatest of biological problems, remote from exact theory, is that of morphogenesis, the mysterious process whereby a nearly undifferentiated droplet of protoplasm, the fertilized ovum, becomes eventually transformed into the marvelous architecture of the multicellular organism. At least a theory of growth as quantitative increase can be developed (cf. pp. 171ff.). This has become a routine method in international fisheries (e.g., Beverton and Holt, 1957). This theory integrates physiology of metabolism and of growth by demonstrating that various types of growth, as encountered in certain groups of animals, depend on metabolic constants. It renders intelligible the equifinality of growth whereby a species-specific final size is attained, even when starting conditions were different or the growth process was interrupted. At least part of morphogenesis is effectuated by so-called relative growth (J. Huxley, 1932), i.e., different growth rates of the various organs. This is a consequence of the competition of these components in the organism for available resources, as can be derived from open system theory (Chapter 7).

Not only the cell, organism, etc., may be considered as open system, but also higher integrations, such as biocoenoses, etc. (cf. Beier, 1962, 1965). The open-system model is particularly evident (and of practical importance) in continuous cell culture as applied in certain technological processes (Malek, 1958, 1964; Brunner, 1967).

These few examples may suffice to indicate briefly the large fields of application of the open-system model. Years ago it was pointed out that the fundamental characteristics of life, metabolism, growth, development, self-regulation, response to stimuli, spontaneous activity, etc., ultimately may be considered as consequences of the fact that the organism is an open system. The theory of such systems, therefore, would be a unifying principle capable of combining diverse and heterogeneous phenomena under the same general concept, and of deriving quantitative laws. I believe this prediction has on the whole proved to be correct and has been testified by numerous investigations.

Behind these facts we may trace the outlines of an even wider generalization. The theory of open systems is part of a *general system theory*. This doctrine is concerned with principles that apply to systems in general, irrespective of the nature of their components and the forces governing them. With general system theory we reach a level where we no longer talk about physical and chemical entities, but discuss wholes of a completely general nature. Yet, certain principles of open systems still hold true and may be applied successfully to wider fields, from ecology, the competition and equilibrium among species, to human economy and other sociological fields.

Open Systems and Cybernetics

Here the important question of the relation of general system theory and cybernetics, of open systems and regulatory mechanisms appears (cf. pp. 160ff.). In the present context a few remarks will suffice.

The basis of the open-system model is the dynamic interaction of its components. The basis of the cybernetic model is the feedback cycle (FIG. 1.1) in which, by way of feedback of information, a desired value (*Sollwert*) is maintained, a target is reached, etc. The theory of open systems is a generalized kinetics and thermodynamics. Cybernetic theory is based on feedback and information. Both models have, in respective fields, been successfully applied. However, one has to be aware of their differences and limitations.

The open-system model in kinetic and thermodynamic formulation does not talk about information. On the other hand, a feedback system is closed thermodynamically and kinetically; it has no metabolism.

In an open system increase of order and decrease of entropy is thermodynamically possible. The magnitude, "information," is defined by an expression formally identical with negative entropy. However, in a closed feedback mechanism information can only decrease, never increase, i.e., information can be transformed into "noise," but not vice versa.

An open system may "actively" tend toward a state of higher organization, i.e., it may pass from a lower to a higher state of order owing to conditions in the system. A feedback mechanism can "reactively" reach a state of higher organization owing to "learning," i.e., information fed into the system.

In summary, the feedback model is preeminently applicable to "secondary" regulations, i.e., regulations based on structural arrangements in the wide sense of the word. Since, however, the structures of the organism are maintained in metabolism and exchange of components, "primary" regulations must evolve from the dynamics in an open system. Increasingly, the organism becomes "mechanized" in the course of development; hence later regulations particularly correspond to feedback mechanisms (homeostasis, goal-directed behavior, etc.).

The open-system model thus represents a fertile working hypothesis permitting new insights, quantitative statements and experimental verification. I would like, however, to mention some important unsolved problems.

Unsolved Problems

At present, we do not have a thermodynamic criterion that would define the steady state in open systems in a similar way as maximum entropy defines equilibrium in closed systems. It was believed for some time that such criterion was provided by minimum entropy production, a statement known as "Prigogine's Theorem." Although it is still taken for granted by some biologists (e.g., Stoward, 1962), it should be emphasized that Prigogine's Theorem, as was well known to its author, applies only under rather restrictive conditions. In particular, it does not define the steady state of chemical reaction systems (Denbigh, 1952; von Bertalanffy, 1953a, 1960b; Foster *et al.*, 1957). A more recent generalization of the theorem of minimum entropy production (Glansdorff and Prigogine, 1964; Prigogine, 1965) encompassing kinetic considerations has still to be evaluated in its consequences.

Another unsolved problem of a fundamental nature originates in a basic paradox of thermodynamics. Eddington called entropy "the arrow of time." As a matter of fact, it is the irreversibility of physical events, expressed by the entropy function, which gives time its direction. Without entropy, i.e., in a universe of completely reversible processes, there would be no difference between past and future. However, the entropy functions do not contain time explicitly. This is true of both the classical entropy function for closed systems by Clausius, and of the generalized function for open systems and irreversible thermodynamics by Prigogine. The only attempt I know of to fill this gap is a further generalization of irreversible thermodynamics by Reik (1953), who attempted to introduce time explicitly into the equations of thermodynamics.

A third problem to be envisaged is the relation between irreversible thermodynamics and information theory. Order is the basis of organization and therefore the most fundamental problem in biology. In a way, order can be measured by negative entropy in the conventional Boltzmann sense. This was shown, e.g., by Schulz (1951) for the nonrandom arrangement of amino acids within a protein chain. Their organization in contrast to hazard arrangement can be measured by a term called chain entropy (*Kettenentropie*). However, there exists a different approach to the problem, i.e., by measurement in terms of yes-or-no decisions, so-called bits, within the framework of information theory. As is well-known, information is defined by a term formally identical with negative entropy, thus indicating a correspondence between the two different theoretical systems of thermodynamics and of information theory. Elaboration of a dictionary, as it were, for translating the language of thermodynamics into that of information theory and vice versa, would seem to be the next step. Obviously, generalized irreversible thermodynamics will have to be employed for this purpose because it is only in open systems that maintenance and elaboration of order do not run contrary to the basic entropy principle.

The Russian biophysicist Trincher (1965) came to the conclusion that the state function, entropy, is not applicable to living systems; he contrasts the entropy principle of physics with biological "principles of adaptation and evolution," expressing an increase of information. Here we have to take into consideration that the entropy principle has a physical basis in the Boltzmann derivation, in statistical mechanics and in the transition toward more probable distributions as is necessary in chance processes; presently, no physical explanation can be given for Trincher's phenomenological principles.

Here we are dealing with fundamental problems which, I believe, "are swept under the carpet" in the present biological creed. Today's synthetic theory of evolution considers evolution to be the result of chance mutations, after a well-known simile (Beadle, 1963), of "typing errors" in the reduplication of the genetic code, which are directed by selection, i.e., the survival of those populations or genotypes that produce the highest number of offspring under existing external conditions. Similarly, the origin of life is explained by a chance appearance of organic compounds (amino acids, nucleic acids, enzymes, ATP, etc.) in a primeval ocean which, by way of selection, formed reproducing units, viruslike forms, protoorganisms, cells, etc.

In contrast to this it should be pointed out that selection, competition and "survival of the fittest" already *presuppose* the existence of self-maintaining systems; they therefore cannot be the *result* of selection. At present we know no physical law which would prescribe that, in a "soup" of organic compounds, open systems, self-maintaining in a state of highest improbability, are formed. And even if such systems are accepted as being "given," there is no law in physics stating that their evolution, on the whole, would proceed in the direction of increasing organization, i.e., improbability. Selection of genotypes with maximum off-spring helps little in this respect. It is hard to understand why, owing to differential reproduction, evolution ever should have gone beyond rabbits, herring or even bacteria, which are un-rivaled in their reproduction rate. Production of local conditions of higher order (and improbability) is physically possible only if "organizational forces" of some kind enter the scene; this is the case in the formation of crystals, where "organizational forces" are represented by valencies, lattice forces, etc. Such organiza-tional forces, however, are explicitly denied when the genome is considered as an accumulation of "typing errors."

Future research will probably have to take into consideration irreversible thermodynamics, the accumulation of information in the genetic code and "organizational laws" in the latter. Pres-ently the genetic code represents the *vocabulary* of hereditary substance, i.e., the nucleotide triplets which "spell" the amino acids of the proteins of an organism. Obviously, there must also exist a *grammar* of the code; the latter cannot, to use a psychi-atric expression, be a word salad, a chance series of unrelated words (nucleotide triplets and corresponding amino acids in the protein molecules). Without such "grammar" the code could at best produce a pile of proteins, but not an organized organism. Certain experiences in genetic regulation indicate the existence of such organization of the hereditary substratum; their effects will have to be studied also in macroscopic laws of evolution (von Bertalanffy, 1949a; Rensch, 1961). I therefore believe that the presently generally accepted "synthetic theory of evolution" is at best a partial truth, not a complete theory. Apart from additional biological research, physical considerations have to be taken into account, in the theory of open systems and its present border-line problems.

Conclusion

The model of the organism as open system has proved useful in the explanation and mathematical formulation of numerous life phenomena; it also leads, as is to be expected in a scientific working hypothesis, to further problems, partly of a fundamental nature. This implies that it is not only of scientific but also of "meta-scientific" importance. The mechanistic concept of nature predominant so far emphasized the resolution of happenings into linear causal chains; a conception of the world as a result of chance events, and a physical and Darwinistic "play of dice" (Einstein); the reduction of biological processes to laws known from inanimate nature. In contrast to this, in the theory of open systems (and its further generalization in general system theory), principles of multivariable interaction (e.g., reaction kinetics, fluxes and forces in irreversible thermodynamics) become apparent, a dynamic organization of processes and a possible expansion of physical laws under consideration of the biological realm. Therefore, these developments form part of a new formulation of the scientific world view.

7 Some Aspects of System Theory in Biology

Introducing the present symposium on Quantitative Biology of Metabolism, the speaker's task, it would seem, is to outline the conceptual framework of the field, illustrating its leading ideas, theories, or—as we may preferably say—the *conceptual constructs* or *models* applied.

According to widespread opinion, there is a fundamental distinction between "observed facts" on the one hand—which are the unquestionable rock bottom of science and should be collected in the greatest possible number and printed in scientific journals —and "mere theory" on the other hand, which is the product of speculation and more or less suspect. I think the first point I should emphasize is that such antithesis does not exist. As a matter of fact, when you take supposedly simple data in our field—say, determination of Qo_2, basal metabolic rates or temperature coefficients—it would take hours to unravel the enormous amount of theoretical presuppositions which are necessary to form these concepts, to arrange suitable experimental designs, to create machines doing the job—and this all is implied in your supposedly raw data of observation. If you have obtained a series of such values, the most "empirical" thing you can do is to present them in a table of mean values and standard deviations. This presupposes the model of a binomial distribution—and with this, the whole theory of probability, a profound and to a large extent unsolved problem of mathematics, philosophy and even metaphysics. If you are lucky, your data can be plotted in a simple fashion, obtaining the graph of a straight line. But considering the unconceivable complexity of processes even in a simple cell, it is little short of a miracle that the simplest possible model—namely, a linear equation between two variables—actually applies in quite a number of cases.

Thus even supposedly unadulterated facts of observation already are interfused with all sorts of conceptual pictures, model concepts, theories or whatever expression you choose. The choice is not whether to remain in the field of data or to theorize; the choice is only between models that are more or less abstract, generalized, near or more remote from direct observation, more or less suitable to represent observed phenomena.

On the other hand, one should not take scientific models too seriously. Kroeber (1952), the great American anthropologist, once made a learned study of ladies' fashions. You know, sometimes skirts go down until they impede the lady in walking; again, up they go to the other possible extreme. Quantitative analysis revealed to Kroeber a secular trend as well as short-period fluctuations in the length of ladies' skirts. This is a perfectly good little law of nature; however, it has little to do with the ultimate reality of nature. I believe a certain amount of intellectual humility, lack of dogmatism, and good humor may go a long way to facilitate otherwise embittered debates about scientific theories and models.

It is in this vein that I am going to discuss four models which are rather fundamental in the field of quantitative metabolism. The models I chose are those of the organism as open system and steady state, of homeostasis, of allometry, and the so-called Bertalanffy model of growth. This is not to say that these models are the most important ones in our field; but they are used rather widely and can illustrate the conceptual framework as well as others.

Open Systems and Steady States

Any modern investigation of metabolism and growth has to take into account that the living organism as well as its components are so-called open systems, i.e., systems maintaining themselves in a continuous exchange of matter with environment (FIG. 7.1). The essential point is that open systems are beyond the limits of conventional physical chemistry in its two main branches, kinetics and thermodynamics. In other terms, conventional kinetics and thermodynamics are not applicable to many

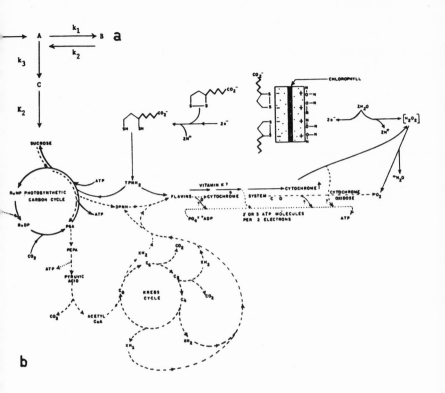

Fig. 7.1. a: Model of a simple open system, showing maintenance of constant concentrations in the steady state, equifinality, adaptation and stimulus-response, etc. The model can be interpreted as a simplified schema for protein synthesis (*A*: amino acids, *B*: protein, *C*: deamination products; k_1: polymerization of amino acids into protein, k_2: depolymerization, k_3: deamination; $k_2 \ll k_1$, energy supply for protein synthesis not indicated). In somewhat modified form, the model is Sprinson & Rittenberg's (1949) for calculation of protein turnover from isotope experiments. (After von Bertalanffy, 1953a).

b: The open system of reaction cycles of photosynthesis in algae. (After Bradley & Calvin, 1957) processes in the living organism; for biophysics—the application of physics to the living organism—an *expansion* of theory is necessary.

The living cell and organism is not a static pattern or machine-like structure consisting of more or less permanent "building materials" in which "energy-yielding materials" from nutrition are broken down to provide the energy requirements for life processes. It is a continuous process in which both so-called building materials as well as energy-yielding substances (*Bau-* and *Betriebsstoffe* of classical physiology) are broken down and re-generated. But this continuous decay and synthesis is so regulated that the cell and organism are maintained approximately constant in a so-called steady state (*Fliessgleichgewicht*, von Bertalanffy). This is one fundamental mystery of living systems; all other characteristics such as metabolism, growth, development, self-regulation, reproduction, stimulus-response, autonomous activity, etc., are ultimately consequences of this basic fact. The organism's being an "open system" is now acknowledged as one of the most fundamental criteria of living systems, at least so far as German science is concerned (e.g., von Bertalanffy, 1942; Zeiger, 1955; Butenandt 1955, 1959).

Before going further, I wish to apologize to the German colleagues for dwelling on matters which are familiar to them, and which I myself have often presented. As Dost (1962a) stated in a recent paper, "our sons already in their premedical examination take account of this matter," i.e., of the theory of open systems in their kinetic and thermodynamic formulations. Remember—to quote but two examples—the presentation of the topic by Blasius (1962) in the new editions of our classic Landois-Rosemann textbook, and Netter in his monumental *Theoretical Biochemistry* (1959). I am sorry to say that the same does not apply to biophysics and physiology in the United States. I have looked in vain into leading American texts even to find the terms, "open system," "steady state" and "irreversible thermodynamics." That is to say, precisely that criterion which fundamentally distinguishes living systems from conventional inorganic ones is generally ignored or bypassed.

Consideration of the living organism as an open system exchanging matter with environment comprises two questions: first, their *statics,* i.e., maintenance of the system in a time-independent state; secondly, their *dynamics,* i.e., changes of the system in time.

The problem can be considered from the viewpoints of kinetics and of thermodynamics.

Detailed discussion of the theory of open systems can be found in the literature (extensive bibliographies in von Bertalanffy 1953a, 1960b). So I shall restrict myself to saying that such systems have remarkable features of which I will mention only a few. One fundamental difference is that closed systems *must* eventually attain a time-independent state of chemical and thermodynamic equilibrium; in contrast, open systems *may* attain, under certain conditions, a time-independent state which is called a steady state, *Fliessgleichgewicht*, using a term which I introduced some twenty years ago. In the steady state, the composition of the system remains constant in spite of continuous exchange of components. Steady states or *Fliessgleichgewichte* are equifinal (Fig. **6.1**); i.e., the same time-independent state may be reached from different initial conditions and in different ways—much in contrast to conventional physical systems where the equilibrium state is determined by the initial conditions. Thus even the simplest open reaction systems show that characteristic which defines biological restitution, regeneration, etc. Furthermore, classical thermodynamics, by definition, is only concerned with closed systems, which do not exchange matter with environment. In order to deal with open systems, an expansion and generalization was necessary which is known as *irreversible thermodynamics*. One of its consequences is elucidation of an old vitalistic puzzle. According to the second principle of thermodynamics, the general direction of physical events is toward states of maximum entropy, probability and molecular disorder, levelling down existing differentiations. In contrast and "violent contradiction" to the second principle (Adams, 1920), living organisms maintain themselves in a fantastically improbable state, preserve their order in spite of continuous irreversible processes and even proceed, in embryonic development and evolution, toward ever higher differentiations. This apparent riddle disappears by the consideration that the classic second principle by definition pertains only to closed systems. In open systems with intake of matter rich in high energy, maintenance of a high degree of order and even advancement toward higher order is thermodynamically permitted.

Living systems are maintained in a more or less rapid exchange, degeneration and regeneration, catabolism and anabolism of their components. The living organism is a hierarchical order of open systems. What imposes as an enduring structure at a certain level, in fact, is maintained by continuous exchange of components of the next lower level. Thus, the multicellular organism maintains itself in and by the exchange of cells, the cell in the exchange of cell structures, these in the exchange of composing chemical compounds, etc. As a general rule, turnover rates are the faster the smaller the components envisaged (Tables 6.1-3). This is a good illustration for the Heraclitean flow in and by which the living organism is maintained.

So much about the statics of open systems. If we take a look at changes of open systems in time, we also find remarkable characteristics. Such changes may occur because the living system initially is in an unstable state and tends toward a steady state; such are, roughly speaking, the phenomena of growth and development. Or else, the steady state may be disturbed by a change in external conditions, a so-called stimulus; and this—again roughly speaking—comprises adaptation and stimulus-response. Here too characteristic differences to closed systems obtain. Closed systems generally tend toward equilibrium states in an asymptotic approach. In contrast, in open systems, phenomena of false start and overshoot may occur (Fig. 6.2). In other terms: If we find overshoot or false start—as is the case in many physiological phenomena—we may expect this to be a process in an open system with certain predictable mathematical characteristics.

As a review of recent work (Chapter 6) shows, the theory of the organism as an open system is a vividly developing field as it should be, considering the basic nature of biological *Fliessgleichgewicht*. The above examples are given because, after the basic investigations by Schönheimer (1947) and his group into the "Dynamic State of Body Constituents" by way of isotope tracers, the field is strangely neglected in American biology which, under the influence of cybernetic concepts, rather has returned to the machine concept of the cell and organism, thereby neglecting the important principles offered by the theory of open systems.

Feedback and Homeostasis

Instead of the theory of open systems, another model construct is more familiar to the American school. It is the concept of feedback regulation, which is basic in cybernetics and was biologically formulated in Cannon's concept of homeostasis (e.g., Wiener, 1948; Wagner, 1954; Mittelstaedt, 1954, 1956; Kment, 1957). We can give it only a brief consideration.

As is generally known, the basic model is a circular process where part of the output is monitored back, as information on the preliminary outcome of the response, into the input (FIG. 7.2a), thus making the system self-regulating; be it in the sense of maintenance of certain variables or of steering toward a desired goal. The first is the case, e.g., in a simple thermostat and in the maintenance of constant temperature and many other parameters in the living organism; the second, in self-steering missiles and proprioceptive control of voluntary movements. More elaborate feedback arrangements in technology and physiology (e.g., FIG. 7.2b) are variations or aggregates of the basic scheme.

Phenomena of regulation following the feedback scheme are of widest distribution in all fields of physiology. Furthermore, the concept appeals to a time when control engineering and automation are flourishing, computers, servomechanisms, etc., are in the center of interest, and the model of the "organism as servomechanism" appeals to the *Zeitgeist* of a mechanized society. Thus the feedback concept sometimes has assumed a monopoly suppressing other equally necessary and fruitful viewpoints: The feedback model is equated with "systems theory" in general (Grodin, 1963; Jones and Gray, 1963; Casey, 1962), or "biophysics" is nearly identified with "computer design and information theory" (Elsasser, 1958, p. 9). It is therefore important to emphasize that feedback systems and "homeostatic" control are a significant but special class of self-regulating systems and phenomena of adaptation (cf. Chapter 6). The following appear to be the essential criteria of feedback control systems:

(1) Regulation is based upon preestablished arrangements ("structures" in a broad sense). This is well expressed by the German term *Regelmechanismen* which makes it clear that the systems envisaged are of the nature of "mechanisms"—in contrast to regulations of a "dynamic" nature resulting from free interplay of forces and mutual interaction between components and tending toward equilibrium or steady states.

(2) Causal trains within the feedback system are linear and uni-

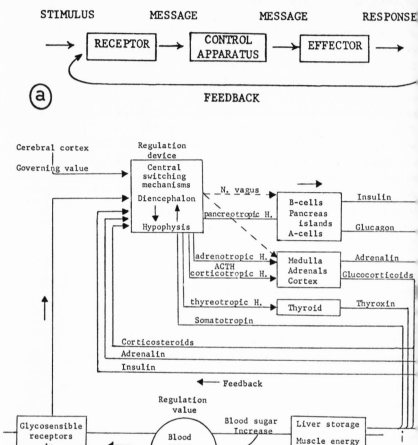

Fig. 7.2. a: Simple feedback scheme. *b*: Homeostatic regulation of the blood sugar level. (After Mittelstaedt, 1954.)

directional. The basic feedback scheme (Fig. **7.2**) is still the classical stimulus-response (*S-R*) scheme, only the feedback loop being added so that causality becomes circular.

(3) Typical feedback or homeostatic phenomena are "open" with respect to incoming information, but "closed" with respect to matter and energy. The concepts of information theory—particularly in the equivalence of information and negative entropy—correspond therefore to "closed" thermodynamics (thermostatics) rather than irreversible thermodynamics of open systems. However, the latter is presupposed if the system (like the living organism) is to be "self-organizing" (Foerster and Zopf, 1962), i.e., is to go toward higher differentiation. As was mentioned above, no synthesis is reached as yet. The cybernetic scheme permits, by way of block diagrams, clarification of many important phenomena of self-regulation in physiology and lends itself to information-theoretical analysis. The open-system scheme permits kinetic and thermodynamic analysis.

Comparison of flow diagrams of feedback (Fig. 7.2) and open systems (Fig. 7.1) intuitively shows the difference. Thus dynamics in open systems and feedback mechanisms are two different model concepts, each in its right in its proper sphere. The open-system model is basically nonmechanistic, and transcends not only conventional thermodynamics, but also one-way causality as is basic in conventional physical theory (cf. Chapter 4). The cybernetic approach retains the Cartesian machine model of the organism, unidirectional causality and closed systems; its novelty lies in the introduction of concepts transcending conventional physics, especially those of information theory. Ultimately, the pair is a modern expression of the ancient antithesis of "process" and "structure"; it will eventually have to be resolved dialectically in some new synthesis.

Physiologically speaking, the feedback model accounts for what may be called "secondary regulations" in metabolism and other fields, i.e., regulations by way of preestablished mechanisms and fixed pathways, as in neurohormonal control. Its mechanistic character makes it particularly applicable in the physiology of organs and organ systems. On the other hand, dynamic interplay of reactions in open systems applies to "primary regulations" such as in cell metabolism (cf. Hess and Chance, 1959) where the more general and primitive open-system regulation obtains.

Allometry and the Surface Rule

Let us now proceed to the third model which is the so-called principle of allometry. As is well known, many phenomena of metabolism, and of biochemistry, morphogenesis, evolution, etc., follow a simple equation:

$$y = bx^\alpha, \tag{7.1}$$

i.e., if a .variable y is plotted logarithmically against another variable x, a straight line results. There are so many cases where this equation applies that examples are unnecessary. Therefore let us look instead at fundamentals. The so-called allometric equation is, in fact, the simplest possible law of relative growth, the term taken in the broadest sense; i.e., increase of one variable, y, with respect to another variable x. We see this immediately by writing the equation in a somewhat different form:

$$\frac{dy}{dt} \cdot \frac{1}{y} : \frac{dx}{dt} \cdot \frac{1}{x} = \text{Rel. Gr. Rate } (y, x) = \alpha. \tag{7.2}$$

As can easily be seen, the allometric equation is a solution of this function which states that the ratio of the relative increase of variable y to that of x is constant. We arrive at the allometric relation in a simple way by considering that any relative growth —only presupposed it is continuous—can generally be expressed by:

$$\text{R. G. R. } (y,x) = F, \tag{7.3}$$

where F is some undefined function of the variables concerned. The simplest hypothesis is that F be a constant, α, and this is the principle of allometry.

However, it is well known that historically the principle of allometry came into physiology in a way very different from the derivation given. It appeared in a much more special form when Sarrus and Rameaux found around 1840 that metabolic rate in animals of different body weight does not increase in proportion to weight, but rather in proportion to surface. This is the origin of the famous surface law of metabolism or law of Rubner, and it is worthwhile to take a look at Rubner's original data of about 1880 (Table 7.1). In dogs of varying weight, metabolic rate decreases if calculated per unit of weight; it remains approximately constant per unit surface, with a daily rate of about 1000 kcal. per square meter. As is well known, the so-called surface law has caused an enormous debate and literature. In fact, Rubner's law is a very special case of the allometric function, y representing

Table 7.1

Metabolism in dogs. (After RUBNER around 1880)

weight in kg	cal. production per kg	cal. production per sq. m body surface
3.1	85.8	1909
6.5	61.2	1073
11.0	57.3	1191
17.7	45.3	1047
19.2	44.6	1141
23.7	40.2	1082
30.4	34.8	984

basal metabolic rate, x body weight, and the exponent α amounting to 2/3.

I believe that the general derivation just mentioned puts the surface law into correct perspective. Endless discussions of some 80 years are overcome when we consider it a special case of allometry, and take the allometric equation for what it really is: a highly simplified, approximate formula which applies to an astonishingly broad range of phenomena, but is neither a dogma nor an explanation for everything. Then we shall expect all sorts of allometric relationships of metabolic measures and body size—with a certain preponderance of surface or 2/3-power functions, considering the fact that many metabolic processes are controlled by surfaces. This is precisely what we find (Table **7.2**). In other words, 2/3 is not a magic number; nor is there anything sacred about the 3/4 power which more recently (Brody, 1945; Kleiber, 1961) has been preferred to the classical surface law. Even the expression: *Gesetz der fortschreitenden Stoffwechselreduktion* (Lehmann, 1956)—law of progressive reduction of metabolic rate—is not in place because there are metabolic processes which do not regress with increasing size.

Furthermore, from this it follows that the dependence of metabolic rates on body size is not invariable as was presupposed by the surface law. It rather can vary, and indeed does vary, especially as a function of (1) the organism or tissue in question; (2) physiological conditions; and (3) experimental factors.

As to the variation of metabolic rate depending on the *organism*

Table 7.2

Equations relating quantitative properties with body weights among mammal
(After ADOLPH 1949; modified)

	regression $\alpha =$		regression $\alpha =$
intake of water (ml/hr)	.88	myoglobin wt (g)	1.31
urine output (ml/hr)	.82	cytochrome wt (g)	.62
urea clearance (ml/hr)	.72	nephra number	.62
inulin clearance (ml/hr)	.77		
creatinine clearance (ml/hr)	.69	diameter renal corp. (cm)	.08
diodrast clearance (ml/hr)	.89	kidneys wt (g)	.85
hippurate clearance (ml/hr)	.80	brain wt (g)	.70
O_2 consum. basal (ml STP/hr)	.734	heart wt (g)	.98
heartbeat duration (hr)	.27	lungs wt (g)	.99
breath duration (hr)	.28	liver wt (g)	.87
ventilation rate (ml/hr)	.74	thyroids wt (g)	.80
		adrenals wt (g)	.92
tidal volume (ml)	1.01	pituitary wt (g)	.76
gut beat duration (hr)	.31	stom. + intes. wt (g)	.94
N total output (g/hr)	.735	blood wt (g)	.99
N endogenous output (g/hr)	.72		
creatinine N output (g/hr)	.90	*Surface law:* $\alpha = .66$ relative to absolute
sulphur output (g/hr)	.74	weight ($y = bw^\alpha$); $- .33$ relative to unit
O_2 consum. liver slices		
(ml STP/hr)	.77	weight $\left(\dfrac{y}{w} = bw^\alpha \right)$
hemoglobin wt (g)	.99	

or *tissue* concerned, I shall give later on examples with respect to total metabolism. Differences in size dependence of Qo_2 in various tissues are shown in Figure 7.3. A similar example is presented in Table 7.3 with respect to comparison of intra- and interspecific allometries. Variations of size-dependence of metabolic rate with *physiological conditions* are demonstrated by data obtained in our laboratory in an important aspect which has been little investigated. The size-dependence of metabolism as expressed in the allometry exponent α varies, depending on whether basal metabolic rate (B.M.R.), resting metabolism, or metabolism in muscular activity is measured. Figure 7.4 shows such variation in rats, comparing basal and nonbasal metabolic rates. Figure 7.5 gives a more extensive comparison in mice, including different degrees of muscular activity. These data confirm Locker's statement (1961a) that with increasing intensity of metabolic rate, α tends to decrease. Variations in the slope of the regression lines are also found in invertebrates when metabolic rates of fasting and nonfasting animals are compared (FIG. 7.6). Variations of α with *experimental conditions* deserve much more attention than usually given. Often the attitude is

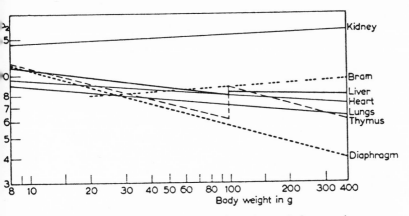

Fig. 7.3. Qo_2 (μlO_2/mg dry wt./hr.) of several rat tissues. Only regression lines are shown in this and the following figures; for complete data see originals. (After von Bertalanffy & Pirozynski, 1953.)

Table 7.3

Intraspecific and interspecific allometry (constants α) in organs of mammals. (After VON BERTALANFFY & PIROZYNSKI 1952)

	rat B. & P.	BRODY	cat	dog various	monkey authors	cattle	horse	adult mammals inter-specific
brain	0.20	0.17		0.25	0.62	0.30	0.24	0.66 0.69 0.58 0.54
heart	0.82	0.80	♂ 0.92 ♀ 0.82	1.00 0.86 0.93	0.69	0.93		0.83 0.82 0.85 0.84 0.98
lungs	0.73	0.75		0.82	0.92		0.58	0.98 0.99
liver	1. Cycle: 1.26 2. Cycle: 0.67	1. Cycle: 1.14 2. Cycle 0.68		0.71		0.70	0.61	0.87 0.88 0.92
kidneys	0.80	0.82	♂ 0.65 ♀ 0.61	0.70			0.66	0.85 0.87 0.76

taken as if Qo_2 were a constant characteristic of the tissue under consideration. This is by no means the case. Variations appear, for example, with different bases of reference, such as fresh weight, dry weight, N-content, etc. (Locker, 1961 b). The simplest

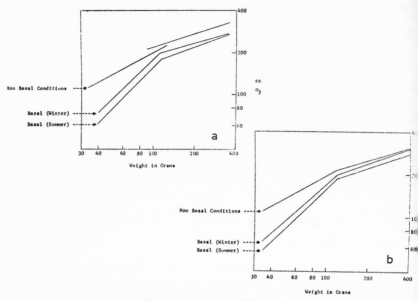

Fig. 7.4. Size dependence of metabolic rates in rat under basal and non-basal conditions. Animals fasted for 18 hrs. prior to experiment (small animals less) ; determinations at 29°–30°C; conditions of muscular rest. A break in the regression lines is assumed at a body weight of 110 gm., corresponding with many physiological changes (cf. *Fig. 7.9*). "Basal Summer" determinations were made with a climatization period of 15–18 hours at thermoneutrality preceding experiment; "Basal Winter" without climatization; "Nonbasal conditions" with 10 hours fasting, followed by a meal 45–60 minutes prior to experiment. *a ♂ , b ♀* (Unpublished data by Racine & von Bertalanffy.)

demonstration is change of the medium. Not only—as every experimenter knows—does the absolute magnitude of Qo_2 vary greatly depending, e.g., on whether saline or medium with metabolites is used; the same is true of size dependence or the parameter α (FIG. **7.7**). Locker's rule, as mentioned previously, again is verified; its confirmations by the experiments summarized in Figures **7.4**, **7.5** and **7.7** are particularly impressive because they were obtained independent of and prior to statement of the rule. The variation of Qo_2 in different media indicates that different partial processes in respiration are measured.

This is the reason why I doubt that total metabolism or B.M.R. can be obtained by so-called summated tissue respiration (Martin and Fuhrmann, 1955). Which Qo_2 of the individual tissues

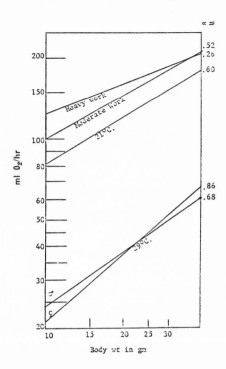

Fig. 7.5. Size dependence of metabolic rates in mice. Determinations at 29° and 21°C: previous fasting and climatization. In the experiments with muscular activity, the scattering of values is considerable owing to the difficulty in keeping the performed work constant. Therefore the qualitative statement that the slope of the regression lines decreases is well established, but no particular significance should be attached to the numerical values of α. (Unpublished data by Racine & von Bertalanffy.)

should be summated? The Qo_2 as obtained, say, in Ringer solution or that obtained with metabolites which may be twice as high? How do the different α's of the various tissues add up to the 2/3 or 3/4 observed in B.M.R. of the entire animal? Similarly, Locker (1962) has shown that also the component processes of Qo_2, such as carbohydrate and fat respiration, may have different regressions.

Before leaving this topic, I would like to make another remark on principle. We have to agree that the allometric equation is, at best, a simplified approximation. Nevertheless, it is more than a convenient way of plotting data. Notwithstanding its simplified

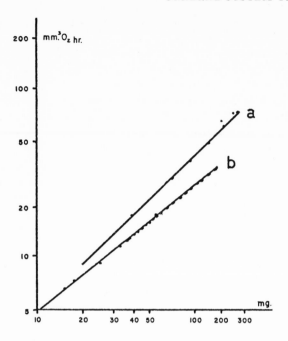

Fig. 7.6. O_2 consumption of larvae of *Tenebrio molitor* (20°C). *a*: larvae fed; *b*: starved for two days. In *b*, Müller's and Teissier's values combined. (After von Bertalanffy & Müller, 1943.)

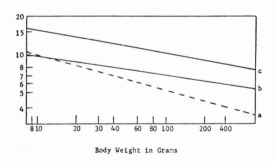

Body Weight in Grams

Fig. 7.7. Size dependence of Qo_2 of diaphragm in different media. a: Krebs-Ringer phosphate solution; *b*: Krebs medium II, type A, with glucose; *c*: Same medium, with glucose and metabolites. (After von Bertalanffy & Estwick, 1953.)

character and mathematical shortcomings, the principle of allometry is an expression of the interdependence, organization and harmonization of physiological processes. Only because processes are harmonized, the organism remains alive and in a steady state. The fact that many processes follow simple allometry, indicates that this is a general rule of the harmonization of processes (Adolph, 1949): "Since so many properties have been found to be adequately interrelated by equations of one form, it seems very unlikely that other properties would be related according to a radically different type of equation. For if they were, they would be incompatible with the properties reviewed."

Furthermore, although we encounter a wide range of values of allometry constants, these certainly are not accidental. At least to a wide extent, they depend on biotechnical principles. It is a truism in engineering that any machine requires changes in proportion to remain functional if it is built in different size, e.g., if a small-scale model is increased to the desired working size. To an extent, it can be understood why certain types of allometry, such as dependence on surface, body mass, etc., obtain in particular cases. The studies by Günther and Guerra (1955) and Guerra and Günther (1957) on biological similarity, the relations of birds' wings (Meunier, 1951), pulse rate (von Bertalanffy, 1960b) and brain weight (von Bertalanffy and Pirozynski, 1952) to body size are examples of functional analysis of allometry which, I believe, will become an important field for further research.

Theory of Animal Growth

The last model I wish to discuss is the model of growth, honorifically called the Bertalanffy equations (von Bertalanffy, 1957b, 1960b); basic ideas go back to the great German physiologist Pütter (1920). Here, too, I am not primarily concerned with details or even the merits and shortcomings of the model; I rather wish to use it to make clear some principles in quantitative metabolism research.

We all know, firstly, that the process of growth is of utmost complexity; and secondly, that there is a large number of formulas on the market which claim satisfactorily to represent observed growth data and curves. The general procedure was that some more or less complex and more or less plausible equation was proposed. Then the investigator sat down to calculate a number of growth curves with that formula, and was satisfied if a sufficient approximation of empirical data was obtained.

Here is a first illusion we have to destroy. It is a mathematical rule of thumb that almost every curve can be approximated if three or more free parameters are permitted—i.e., if an equation contains three or more so-called constants that cannot be verified otherwise. This is true quite irrespective of the particular form of the equation chosen; the simplest equation to be applied is a power series ($y = \alpha_0 + \alpha_1 x + \alpha_2 x^2 + \ldots$) developed to, say, the cubic term. Such calculation is a mere mathematical exercise. Closer approximation can always be obtained by permitting further terms.

The consequence is that curve-fitting may be an indoor sport and useful for purposes of interpolation and extrapolation. However, approximation of empirical data is not a verification of particular mathematical expressions used. We can speak of verification and of equations representing a theory only if (1) the parameters occurring can be confirmed by independent experiment; and if (2) predictions of yet unobserved facts can be derived from the theory. It is in this sense that I am going to discuss the so-called Bertalanffy growth equations because, to the best of my knowledge, they are the only ones in the field which try to meet the specifications just mentioned.

The argument is very simple. If an organism is an open system, its increase or growth rate (G.R.) may, quite generally, be expressed by a balance equation of the form:

$$\frac{dw}{dt} = \text{G.R.} = \text{Synth.} - \text{Deg.} + \ldots, \qquad (7.4)$$

i.e., growth in weight is represented by the difference between processes of synthesis and degeneration of its building materials, plus any number of indeterminate factors that may influence the process. Without loss of generality, we may further assume that the terms are some undefined functions of the variables concerned:

$$\text{G.R.} = f_1(w, t) - f_2(w, t) + \ldots. \qquad (7.5)$$

Now we see immediately that time t should not enter into the equation. For at least some growth processes are equifinal, i.e., the same final values can be reached at different times (FIG. **6.1**). Even without strict mathe.natical proof, we can see intuitively that this would not be possible if growth rate directly depends on time; for if this were the case, different growth rates could not occur at given times as is sometimes the case.

Consequently, the terms envisaged will be functions of body mass present:

$$\text{G.R.} = f_1(w) - f_2(w), \tag{7.6}$$

if we tentatively limit the consideration to the simplest open-system scheme. The simplest assumption we can make is that the terms are power functions of body mass. And, indeed, we know empirically that quite generally the size dependence of physiological processes can well be approximated by allometric expressions. Then we have:

$$\frac{dw}{dt} = \eta w^n - \kappa w^m \tag{7.7}$$

where η and κ are constants of anabolism and catabolism, respectively, corresponding to the general structure of allometric equations.

Mathematical considerations show furthermore that smaller deviations of the exponent m from unity do not much influence the shape of the curves obtained. Thus, for further simplification let us put $m = 1$. This makes things much easier mathematically, and appears to be justified physiologically, since physiological experience—limited it is true—seems to show that catabolism of building materials, especially proteins, is roughly proportional to body mass present.

Now let us make a big leap. Synthesis of building materials needs energy which, in aerobic animals, is provided by processes of cell respiration and ultimately the ATP system. Let us assume there are correlations between energy metabolism of an animal and its anabolic processes. This is plausible insofar as energy metabolism must, in one way or the other, provide the energies that are required for synthesis of body components. We therefore insert for size dependence of anabolism that of metabolic rates ($n = \alpha$) and arrive at the simple equation:

$$\frac{dw}{dt} = \eta w^\alpha - \kappa w. \tag{7.8}$$

The solution of this equation is:

$$w = \left\{ \frac{\eta}{\kappa} - \left(\frac{\eta}{\kappa} - w_0^{(1-\alpha)} \right) e^{-(1-\alpha)\kappa t} \right\}^{1/1-\alpha} \qquad (7.9)$$

with w_0 = weight at time $t = 0$.

Empirically, we find that resting metabolism of many animals is surface-dependent, i.e., that they follow Rubner's rule. In this case, we set $\alpha = 2/3$. There are other animals where it is directly dependent on body mass, and then $\alpha = 1$. Finally, cases are found where metabolic rate is in between surface and mass proportionality, that is $2/3 < \alpha < 1$. Let us tentatively refer to these differences in size-dependence of metabolic rate as "metabolic types."

Now if we insert the different values for α into our basic equation, we easily see that they yield very different curves of growth. Let us refer to them as "growth types." These are summarized in Table 7.4; corresponding graphs, showing the differences in metabolic behavior and concomitant differences of growth curves, are presented in Figure 7.8. Detailed discussions of the theory have been given elsewhere. It has been shown that the above derivations apply in many cases; no less than fourteen different arguments in verification of the theory can be presented (Table 7.5, Figs. 7.9, 7.10). We shall limit the present discussion to a few remarks on principle.

All parameters of the growth equations are verifiable experimentally. α, the size dependence of metabolic rate, determines the shape of the growth curve. This correlation has been confirmed in a wide range of cases, as seen in Table 7.4 κ, constant of catabolism, can in first approximation be identified with turnover of total protein (r) as determined by isotope tracers and other techniques. For example, from the growth curves catabolic rates of 0.045/day for the rat, and 1.165 g. protein/kg. body wt./day for man were calculated (von Bertalanffy, 1938). Determinations of protein catabolism then available did not agree with these predictions: protein loss determined by minimum N-excretion was 0.00282/day for the rat after Terroine, and some 0.4–0.6 g. protein/kg. body wt./day for man, according to the conceptions then prevailing in physiology (von Bertalanffy, 1942, pp. 180ff., 186–188). It was therefore a striking confirmation

Table 7.4

Metabolic types and growth types. w, l: Weight, length at time t; w_0, l_0: initial weight, length; w^*, l^*: final weight, length; η, k: constants of anabolism and catabolism.
(After von Bertalanffy 1942)

Metabolic type	Growth type	Growth equations	Examples
I. Respiration surface-proportional	(a) Linear growth curve: attaining *without inflexion* a steady state. (b) Weight growth curve: *sigmoid*, attaining, with inflexion at c. 1/3 of final weight, a steady state	$dw/dt = \eta w^{2/3} - \varkappa w$ a) $l = l^* - (l^* - l_0)e^{-\varkappa t/3}$ b) $w = [\sqrt[3]{w^*} - (\sqrt[3]{w^*} - w_0)e^{-\varkappa t/3}]^3$	Lamellibranchs, fish, mammals
II. Respiration weight-proportional	Linear and weight growth curves *exponential*, no steady state attained, but growth intercepted by metamorphosis or seasonal cycles	$dw/dt = \eta w - \varkappa w = cw$ a) $l = l_0 e^{ct/3}$ b) $w = w_0 e^{ct}$	Insect larvae, Orthoptera, Helicidae
III. Respiration intermediate between surface- and weight-proportionality	(a) Linear growth curve: attaining *with inflexion* a steady state. (b) Weight growth curve: *sigmoid*, similar to I (b)	$dw/dt = \eta w^a - \varkappa w;$ $^{2/3} < n < 1$ $dl/dt = \dfrac{\eta'}{3} \cdot {}^3 w^{a-2} \cdot \dfrac{\varkappa}{3} l$	Planorbidae

of the theory when later on determinations using the isotope method (Sprinson and Rittenberg, 1949, Table **6.2**) yielded turnover rates of total protein (r) of 0.04/day for the rat, and of 1.3 g. protein/kg. body wt./day for man in an amazing agreement between predicted and experimental values. It may be noted in passing that an estimate of the turnover time of the human organism similar to that found in isotope experiments ($r \approx 0.009$, $t \approx 110$ days) can be obtained in different ways, e.g., also from calorie loss in starvation ($t = 100$ days: Dost, 1962a). η, constant of anabolism, is dimensionally complex. It can, however, be checked by comparison of growth curves of related organisms: according to theory, the ratio of metabolic rates should correspond to the ratio of η's of the animals concerned. This also has been confirmed (Fig. **7.10**).

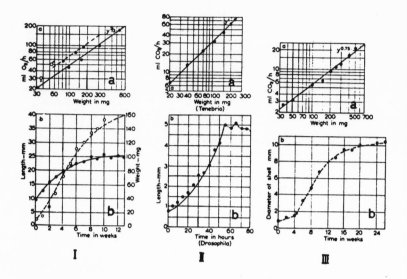

Fig. 7.8. Metabolic and growth types. Type I: *Lebistes reticulatus*; type II: insect larvae; type III: *Planorbis sp. a*: dependence of metabolic rate on body size; *b*: growth curves. (After von Bertalanffy, 1942.)

The theory, therefore, fulfills the first postulate indicated above, i.e., verification of calculated parameters in independent experiments. As has been shown elsewhere, it also fulfills the second postulate: Predictions from the theory were made which came as "surprises," i.e., were unknown at the time, but later on confirmed.

Discussion of some typical objections is in place because it may contribute to better understanding of mathematical models in general.

(1) The main reproach against models and laws for physiological phenomena is that of "oversimplification." In a process such as animal growth there is, at the level of cells, a microcosm of innumerable processes of chemical and physical nature: all the reactions in intermediary metabolism as well as factors like cell permeability, diffusion, active transport and innumerable others.

Table 7.5

Growth of *Acipenser stellatus* (After VON BERTALANFFY 1942)

time in years	length in cm observed	calculated	k
1	21.1	21.1	
2	32.0	34.3	0.062
3	42.3	41.5	0.062
4	51.4	50.8	0.061
5	60.1	59.5	0.061
6	68.0	67.8	0.061
7	75.3	75.5	0.060
8	82.3	82.8	0.060
9	89.0	89.7	0.059
10	95.3	96.2	0.059
11	101.6	102.3	0.059
12	107.6	108.0	0.060
13	112.7	113.4	0.059
14	117.7	118.5	0.059
15	122.2	122.5	0.058
16	126.5	127.9	0.059
17	130.9	132.2	0.059
18	135.3	136.2	0.059
19	140.2	140.0	0.060
20	145.0	143.5	0.061
21	148.6	146.9	0.061
22	152.0	150.0	0.061

Growth equation: $l = 201.1 - (201.1 - 21.1)^{-0.06t}$. Owing to the regularity of growth curves, the BERTALANFFY equations are most suitable for calculation of growth in fish. In this example, the growth constant k ($= \kappa/3$) was calculated in a way similar to calculation of reaction constants in chemical reactions. Variations of this parameter are minimal, so showing the adequacy of the equation.

On the level of organs, each tissue behaves differently with respect to cell renewal and growth; besides multiplication of cells, formation of intercellular substances is included. The organism as a whole changes in composition, with alterations of the content in protein, deposition of fat or simple intake of water; the specific weight of organs changes, not to speak of morphogenesis and differentiation which presently elude mathematical formulation. Is not any simple model and formula a sort of rape of nature, pressing reality into a Procrustean bed and recklessly cutting off what doesn't fit into the mold? The answer is that science in

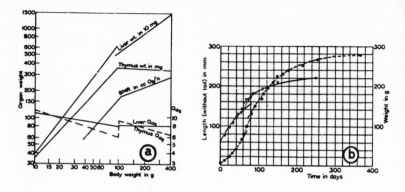

Fig. 7.9. Calculation of growth of the white rat. Many physiological processes in the rat show discontinuities at about 100 gm. body weight, i.e., in the prepubertal stage (*a*). Such "cycle" also appears in metabolism (*Fig. 7.4*), metabolic rates in animals under 100 gm. increasing more, and in animals above this size much less than would correspond to the surface rule. However, if regression is calculated over the whole weight range, a value near 2/3 results as gross average. Hence, in the calculation of the growth curve (1) two "cycles" separated at $\approx$ 100 gm. should appear, and (2) in first approximation, rat growth should be calculable with the equations of "Type I", i.e. $\alpha \approx 2/3$. Calculation of growth data made previous to the physiological determinations (*b*) verifies both expectations. The catabolic constant (κ) results, for the second (postpubertal) cycle, as $\kappa_{calc.} \approx 0.045$/day, in close correspondence with protein turnover determined by isotope tracers ($r=0.04$/day). (After von Bertalanffy, 1960b.)

general consists to a large extent of oversimplifications in the models it uses. These are an aspect of the idealization taking place in every law and model of science. Already Galileo's student, Torricelli, bluntly stated that if balls of stone, of metal, etc., do not follow the law, it is just too bad for them. Bohr's model of the atom was one of the most arbitrary simplifications ever conceived—but nevertheless became a cornerstone of modern physics. Oversimplifications progressively corrected in subsequent development are the most potent or indeed the only means toward conceptual mastery of nature. In our particular case it is not quite correct to speak of oversimplification. What is involved are rather *balance equations* over many complex and partly unknown processes. The legitimacy of such balance expressions is established by routine practice. For example, if we speak of

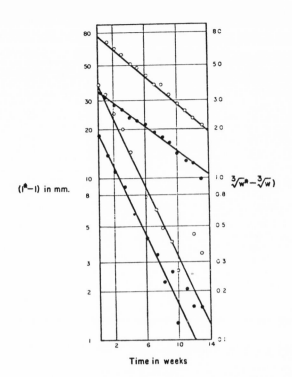

Fig. 7.10. Growth of *Lebistes reticulatus*. Upper lines: ♂, lower lines: ♀ ° weight, ● length. In the Guppy, growth in females and males shows considerable difference, the females reaching a multiple of body weight of the males. Data are logarithmically plotted according to the integral of Equation *7.8*; the close fit shows that the growth curves are correctly reproduced. The growth equations so obtained give a ratio of 1:1.5 for the anabolic constants η in females and males. According to theory, metabolic rates in females and males should stand in the same ratio, 1:1.5 as is actually found (*Fig. 7.8*,I) . (After von Bertalanffy, 1938, 1960b.)

B.M.R.—and are, in fact, able to establish quantitative relationships such as the "surface law"—it is balances we express which nevertheless are important both theoretically and practically (e.g., diagnostic use of B.M.R.). The regularities so observed cannot be refuted by "general considerations" of oversimplification, but only empirically and by offering better explanations. It would be easy to make the growth model seemingly more realistic and to improve fitting of data, by introducing a few more parameters.

However, the gain is spurious as long as these parameters cannot be checked experimentally; and for the reasons mentioned, a closer fit of data tells nothing about the merits of a particular formula if the number of "free constants" is increased.

(2) Another question is the choice of parameters. It has been noted above that metabolic rate under basal and nonbasal conditions changes not only in magnitude but also with respect to allometry expressing its relation to body size. What is the justification of taking "resting metabolism" as standard and to range various species into "metabolic" and "growth types" accordingly? The answer is that among available measures of metabolism— none of them ideal—resting metabolism appears to approach best those natural conditions which prevail during growth. The B.M.R. standard (i.e., thermoneutrality of environment, fasting and muscular rest) makes the values so determined a laboratory artifact, because at least the first condition is unnatural; although it is most useful because B.M.R. values show the least dispersion. In cold-blooded animals, B.M.R. cannot be used as standard because there is no condition of thermoneutrality, and the fasting condition often cannot be exactly established. Activity metabolism, on the other hand, changes with the amount of muscular action (FIG. 7.4), and the growing animal is not under conditions of hard muscular work all the time. Hence resting metabolic rate is comparatively the best approximation to the natural state; and choice of this parameter leads to a useful theory.

(3) The most important criticism becomes apparent from the above discussion. It was said that there appear to be so-called metabolic types and growth types and correlations between both. However, earlier it has been emphasized that the parameters implied, especially the relation of metabolic rate to body size expressed in the exponent α, can be altered and shifted with experimental conditions (FIGS. 7.4–7.7). Similarly, also growth curves are not fixed. Experiments on the rat have shown that the shape of the growth curve, including location and existence of a point of inflection, can be changed by different nutrition (L. Zucker et al., 1941a, 1941b, 1942; T.F. Zucker et al., 1941; Dunn et al., 1947; Mayer, 1948). None of the characteristics is rigid— and, incidentally, within my own biological concepts, I would be the last to presuppose rigidity in the dynamic order of physiological processes. According to my whole biological outlook,

I am rather committed to the ancient Heraclitean concept that what is permanent is only the law and order of change.

However, the apparent contradiction can well be resolved when we remain faithful to the spirit of the theory. What is really invariable is the organization of processes expressed by certain relationships. This is what the theory states and experiments show, namely, that there are *functional relationships* between certain metabolic and growth parameters. This does not imply that the parameters themselves are unchangeable—and the experiments show that they are not. Hence, without loss of generality, we may understand "metabolic" and "growth types" as ideal cases observable under certain conditions, rather than as rigid species characteristics. "Metabolic" and "growth types" appear in the respective groups of animals if certain standard conditions are met. However, it is clearly incorrect that "the reduction of metabolic rates is a fundamental magnitude, not changing in different external conditions" (Lehmann, 1956). Under natural or experimental conditions, the relationships can be shifted, and then a corresponding alteration of growth curves should take place. There are indications that this is actually the case; it is a clear-cut problem for further investigation.

A case to the point are seasonal changes. Berg (1959, 1961), while in general confirming previous data, found that the size-metabolism relation varies seasonally in snails: "Thus the relation, oxygen consumption to body size, is not a fixed, unchangeable quantity characteristic of all species as supposed by Bertalanffy. . . . If (Bertalanffy's theory) were true, then the observed seasonal variation in metabolic type would imply a seasonal variation in the type of growth rate."

As a matter of fact, precisely this has been found in our laboratory long ago (von Bertalanffy and Müller, 1943). Seasonal variations of metabolic rate in snails have been described (FIG. 7.11*a*). But correspondingly, also the growth curve (exponential in this case because these snails belong to "Type II") shows breaks and cycles (FIG. 7.11*b*). Therefore, this certainly is a problem deserving more detailed investigation; however, the data available are a hint toward confirmation rather than refutation of the theory.

I would have been much surprised, indeed suspicious, if this first crude model would have provided a conclusive theory. Such

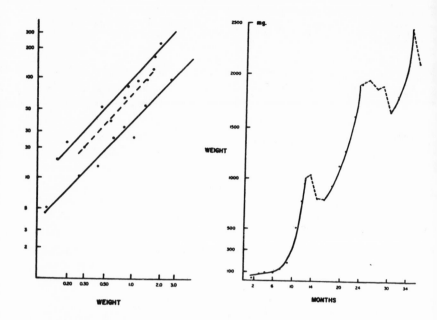

Fig. 7.11. Metabolism and growth in land snails. *a*: Seasonal variations in metabolic rates. The regression lines show, from bottom to top, resting metabolism of *Cepaea vindobonensis* inactive shortly after hibernation at 20°C, same at 28°C, and in activity period at 20°C. (Weight in gm.) Other conditions being equal, resting metabolism is considerably higher in the active compared to the inactive season.

b: Growth in a related species *(Eulota fruticum)*. The growth curve is exponential (Type II with α ≈ 1), but shows seasonal fluctuations. (After von Bertalanffy & Müller, 1943)

things just do not happen, as is witnessed by many examples from history of science. Mendel's laws were the beginnings of genetics but—with linkage, crossing-over, position effect and what not—it is only a minute part of genetic experience that is described by the classical laws. Galileo's law is the beginning of physics, but only highly idealized cases—such as bodies falling in vacuo—actually follow the simple law. It is a long way from Bohr's simple model of the hydrogen atom to present atomic physics, and so on. It would be fantastically improbable if this were different with a proposed model of growth. The most we can say about it is that it is backed by a considerable amount of experimental evidence, has proved to have explanatory and predictive capacities, and offers clear-cut problems for further research.

It is obvious that the theory has been developed for a limited number of cases only, owing to the limited number of good data and the time-consuming nature both of observation and calculation of growth. Hemmingsen (1960) has made this clear: "With n varying as much as the examples show, within any group with allegedly (or at least first allegedly) uniform growth type, it seems impossible to accept Bertalanffy's generalizations unless a statistically significant correlation between n and growth type can be demonstrated on a much larger number of examples than the few ones which Bertalanffy has repeatedly published." I entirely agree with this criticism; many more data would be desirable, although one should not cavalierly bypass those offered in confirmation of the theory, even if they are some 20 years old. I would amend Hemmingsen's criticism by suggesting reexamination on a broader basis. This should include at least the following items: analysis of a large number of growth data, now made possible by electronic computers; concurrent determination of size-dependence of resting metabolism (constant α) in these cases; determinations of protein catabolism (constant κ); determination, in related species, of the ratios between allometry exponents of metabolic rates and the theoretically identical ratios of the anabolic constants (η). These are all interesting and somewhat neglected research problems; and if the model does no more than bring them to the fore, it has proved its usefulness.

Such investigation may bring additional confirmation of the model; it may lead to its modification and elaboration by taking into account additional factors; or it may lead to abandoning the model altogether and replacing it with a better one. If the latter should happen, I would be in no way disappointed. This is exactly what models are for—to serve as working hypotheses for further research.

What I have tried to show in the models discussed are general ways of analysis of quantitative data. I wanted to make clear both the usefulness and the limitations of such models. Any model should be investigated according to its merit with a view at the explanations and predictions it is able to provide. General criticism does not help, and the decision whether or not a model is suitable, exclusively rests with facts of observation and experiment. On the other hand, no model should be taken as conclusive; at best it is an approximation to be progressively worked out and corrected. In close interaction between experiment and conceptualization, but not in confinement to experimentation or construction of purely speculative models, lies the further development of a field like quantitative biology of metabolism.

Summary

1. The theories of open systems, feedback, allometry and growth according to Bertalanffy are reviewed with respect to their experimental applications.

2. The models of both open system and feedback apply to a wide range of phenomena in physiology, and represent essential expansions of physical theory. The two conceptions should be clearly distinguished; the feedback model (homeostasis) should not be considered a cover-all for physiological regulation in general or identified with "systems theory."

3. The allometric equation represents the simplest possible relation between body size and metabolic processes. It is of a wide applicability and expresses the harmonization of processes in living systems. However, there is no "surface" or "3/4-power law" or "law of progressive reduction of metabolic rates." The allometric relationship greatly varies in physiological phenomena.

4. Variations of the relation between body size and metabolic rate may occur (a) in different tissues or in different species; (b) due to changes of physiological conditions; (c) due to different experimental designs. Among the conditions altering this relation are such factors as physiological activities, sex, season, previous acclimation, etc.

5. The size-dependence of total metabolism in mammals is different under basal conditions, in a nonthermoneutral environment, and under conditions of muscular activity. The variations follow Locker's rule, i.e., with an absolute increase of metabolic rate (expressed by the constant b of the allometric equation), regression with respect to body size (expressed by the slope of the allometric line, α) tends to decrease.

6. The growth equations after Bertalanffy represent a highly simplified model which, however, covers many phenomena and regularities found in the physiology of metabolism and growth. The parameters occuring in these equations have been verified by physiological experiments in many cases.

7. In view of the changes of the size-metabolism relation mentioned under (5), Bertalanffy's so-called metabolic and growth types should

be considered as ideal cases realizable under certain standard conditions, rather than as invariable characteristics of the species or group of species concerned.

8. Seasonal variations of metabolic rates and growth rates seem to show correspondence.

9. Urgent problems for further research with respect to each of the basic models are outlined.

8 The System Concept in the Sciences of Man

The Organismic Revolution

In a famous passage of his *Critique of Practical Reason*, Kant stated that there are two things that fill him with indescribable awe—the starry sky above him and the moral law within him. Kant's time was the height of German classicism. Within a few decades before and after 1800 the great German poets, writers and philosophers were clustered, and Kant's philosophy was the culminating synthesis of physical science as it had developed since Galileo and Newton.

Pondering Kant's statement, we wonder. Among the things he could have found objects of awe, he might well have included a third. Kant did not mention *life*—in its aspects both as the miraculous organization of the living organism and as the microcosm of mind which comprehends the physical universe.

It is not difficult to understand Kant's omission. Physics was nearing one of its culminating points to which Kant himself, in his work on the origin of the solar system, had contributed; the moral law had a long history in the Greek and Judeo-Christian tradition. In contrast, the development of the sciences of biology and psychology had scarcely begun.

The 180 years or so since Kant's writing have seen the Industrial Revolution and, in the near past, the Atomic Revolution, the Revolution of Automation and the Conquest of Space. But there appears to be a break. The breathtaking technological development and the affluent society, realized at least in some parts of the globe, have left us with anxiety and meaninglessness. Physics, with all its stupendous modern insights, is not the crystal-clear structure Kant believed it to be. Kant's moral imperative, even if not eroded, would be much too simple for a complex world. Even apart from the menace of physical annihilation, there is the feeling that our world vision and our system of values are breaking down in the advent of Nihilism which Nietzsche prophetically forecast at the turn of our century.

Considered in the light of history, our technology and even our society are based on a physicalistic world picture which found an early synthesis in Kant's work. Physics is still the paragon of science, the basis of our idea of society and our image of man.

In the meanwhile, however, new sciences have arisen—the life, behavioral, and social sciences. They demand their place in a modern world view, and should be able to contribute to a basic reorientation. Less advertised than the contemporary revolutions in technology but equally pregnant of future possibilities is a revolution based on modern developments in biological and behavioral science. For short, it may be called the *Organismic Revolution*. Its core is the notion of *system*—apparently a pale, abstract and empty concept which nevertheless is full of hidden meaning, ferment and explosive potentialities.

The bearing of this new conception can be epitomized in a short statement. The 19th and first half of the 20th century conceived of the *world as chaos*. Chaos was the oft-quoted blind play of atoms which, in mechanistic and positivistic philosophy, appeared to represent ultimate reality, with life as an accidental product of physical processes, and mind as an epiphenomenon. It was chaos when, in the current theory of evolution, the living world appeared a product of chance, the outcome of random mutations and survival in the mill of natural selection. In the same sense, human personality, in the theories of behaviorism as well as of psychoanalysis, was considered a chance product of nature and nurture, of a mixture of genes and an accidental sequence of events from early childhood to maturity.

Now we are looking for another basic outlook on the world—

the world as organization. Such a conception—if it can be substantiated—would indeed change the basic categories upon which scientific thought rests, and profoundly influence practical attitudes.

This trend is marked by the emergence of a bundle of new disciplines such as cybernetics, information theory, general system theory, theories of games, of decisions, of queuing and others; in practical application, systems analysis, systems engineering, operations research, etc. They are different in basic assumptions, mathematical techniques and aims, and they are often unsatisfactory and sometimes contradictory. They agree, however, in being concerned, in one way or the other, with "systems," "wholes" or "organization"; and in their totality, they herald a new approach.

The Image of Man in Contemporary Thought

What can these developments contribute toward the Sciences of Man? The unsatisfactory status of contemporary psychological theory is common knowledge. It seems a hodgepodge of contradicting theories ranging from behaviorism, which sees no difference between human behavior and that of laboratory rats (and, more important, engineers pattern human behavior after the model of rat behavior), to existentialism, for which the human situation is beyond scientific understanding. The variety of conceptions and approaches would be quite healthy, were it not for one disturbing fact. All these theories share one "image of man" which originated in the physical-technological universe; which is taken for granted by otherwise antagonistic theories such as those of behaviorism, computer models of cognitive processes and behavior, psychoanalysis and even existentialism; and which is demonstrably false. This is the robot model of human behavior.

It is, of course, true that there are a considerable number of trends toward new conceptions, urged on by the insight that the robot model is theoretically inadequate in view of empirical fact and is practically dangerous in its application to "behavioral engineering." Nevertheless, while robotic concepts are frequently denounced overtly and covertly, they remain dominant in psychological research, theory, and engineering. They therefore deserve brief consideration even now.

One leading concept is the *stimulus-response scheme*, or S-R scheme for short. Behavior, animal and human, is considered to be response to stimuli coming from outside. In part, stimulus-response is based upon inherited neural mechanisms, as in reflexes and instinctive behavior. The more important part, so far as human behavior is concerned, are acquired or conditioned responses. This may be classical conditioning by way of repetition of the sequence of conditional and unconditional stimuli according to Pavlov. It may be operant conditioning by reinforcement of successful responses according to Skinner. It may be early childhood experience according to Freud, beginning with toilet training and other procedures whereby socially acceptable behavior is reinforced, but psychopathological complexes may also be formed. This, then, dominates psychological engineering. Scholastic learning is best carried through by teaching machines constructed according to the Skinnerian principles. Conditioning with psychoanalytic background keeps the wheels of free enterprise going. Advertising, motivation research, radio and television are ways of conditioning or programming the human machine so that it buys what it should: the washing powder wrapped in the most brilliant color, the biggest refrigerator as symbol of the maternal womb, or the political candidate commanding the most efficient party machine.

The point is that the rules found by learning theorists in animal experiments are supposed to cover the total of human behavior. To Skinner, for example, the "verbal behavior" of the child is supposedly acquired in the same process of operant conditioning as Skinner's rats and pigeons learn their little tricks by being gratified with small parcels of food for correct responses. As a witty critic (Chomsky, 1959) noted, parents supposedly teach their child to walk and to speak because their teaching behavior is reinforced by gratification, probably so that the child later on may make some money by delivering newspapers, or can call his parents to the telephone. More sophisticated versions of the scheme do not alter its essence.

A second principle is that of *environmentalism* which states, in accordance with the S-R scheme, that behavior and personality are shaped by outside influences. The famous expression is that by Watson: Give me a bunch of kids, (said the founder of behaviorism), taken as they come—and I will make them doctors, lawyers, merchant men, beggars and thieves, solely by the power of conditioning. It is the same principle when psychoanalysis says that personality is formed by early childhood experience, especially of a sexual nature. In more general formulation, the human brain is a computer that can be programmed at will. The practical consequence is that human beings are born not only with equal rights but with equal capabilities. Hence our almost pathological concern with the abnormal, the mentally ill and outright criminal who, by suitable reconditioning, should be brought back into the flock—often to the detriment of consideration given the healthy, normal or superior. Hence also the belief that money buys everything: when the Russians build better space vehicles, a few more billions spent on education will produce the crop of young Einsteins needed for closing the gap.

The third is *the equilibrium principle*. In Freudian formulation, this is the "principle of stability": the basic function of the mental apparatus consists in maintaining homeostatic equilibrium. Behavior essentially is reduction of tensions, particularly those of a sexual nature. Hence, let them release their tensions by way of promiscuity and other tension reduction, and you will have normal and satisfied human beings.

Fourthly, behavior is governed by the *principle of economy*. It is *utilitarian* and should be carried through in the most economic way, that is, at minimum expense of mental or vital energy. In practice, the economic principle amounts to the postulate of minimum demands: for example, reduce scholastic demands to the minimum necessary to become an executive, electronics engineer or plumber—otherwise you warp personality, create tensions, and make an unhappy being.

The present crisis of psychology (which, however, has already lasted for some 30 years) can be summarized as the slow erosion of the robot model of man which up to recent years dominated psychology, particularly in the United States.

Two points deserve to be reemphasized. First, the model of man as robot was germane to all fields of psychology and psychopathology, and to theories and systems otherwise different or antagonistic: to the S-R theory of behavior; to cognitive theory in what has been called the "dogma of immaculate perception," i.e. the organism as a passive receptor of stimuli; to learning theories, Pavlovian, Skinnerian, or with intervening variables; to diverse personality theories; to behaviorism, psychoanalysis, cybernetic concepts in neurophysiology and psychology, and so on. Furthermore, "man as robot" was both expression and motor force of the *zeitgeist* of a mechanized and commercialized society; it helped to make psychology the handmaiden of pecuniary and political interests. It is the goal of manipulating psychology to make humans ever more into robots or automata, this being engineered by mechanized learning, advertising techniques, mass media, motivation research and brainwashing.

Nevertheless, these basic presuppositions are spurious. That is to say, conditioning and learning theories correctly describe an important part or aspect of human behavior, but taken as a nothing-but theory they become ostensibly false and self-defeating in their application. The image of man as robot is metaphysics or myth, and its persuasiveness rests only in the fact that it so closely corresponds to the mythology of mass society, the glorification of the machine, and the profit motive as sole motor of progress.

Unbiased observation easily shows the spuriousness of these basic assumptions. The S-R scheme leaves out the large part of behavior which is expression of spontaneous activities such as play, exploratory behavior and any form of creativity. Environmentalism is refuted by the elementary fact that not even fruit flies or Pavlovian dogs are equal, as any student of heredity or behavior should know. Biologically, life is not maintenance or restoration of equilibrium but is essentially maintenance of disequilibria, as the doctrine of the organism as open system reveals. Reaching equilibrium means death and consequent decay. Psychologically, behavior not only tends to release tensions but also builds up tensions; if this stops, the patient is a decaying mental corpse in the same way a living organism becomes a body in decay when tensions and forces keeping it from equilibrium have stopped. Juvenile delinquents who commit crime for fun, a new psychopathology resulting from too much leisure, the fifty percent mental cases in our hospitals—all this is proof that the scheme of adaptation, adjustment, conformity, psychological and social equilibrium doesn't work. There is a wide range of behavior—and, presumably also of evolution—which cannot be reduced to utilitarian principles of adaptation of the individual and survival of the species. Greek sculpture, Renaissance painting. German music—indeed, any aspect of culture—

has nothing to do with utility, or with the better survival of individuals or nations. Mr. Babbitt is in every utilitarian respect better off than Beethoven or Michelangelo.

Also the principle of stress, so often invoked in psychology, psychiatry and psychosomatics, needs some reevaluation. As everything in the world, stress too is an ambivalent thing. Stress is not only a danger to life to be controlled and neutralized by adaptive mechanisms; it also creates higher life. If life, after disturbance from outside, had simply returned to the so-called homeostatic equilibrium, it would never have progressed beyond the amoeba which, after all, is the best adapted creature in the world—it has survived billions of years from the primeval ocean to the present day. Michelangelo, implementing the precepts of psychology, should have followed his father's request and gone in the wool trade, thus sparing himself lifelong anguish although leaving the Sistine Chapel unadorned.

Selye wrote: "The secret of health and happiness lies in successful adaptation to the ever-changing conditions of the globe; the penalties for failure in this great process of adaptation are disease and unhappiness" (1956, p. VII). He speaks for the worldly-wise and in a sense he is correct. But, taken literally, he would negate all creative activity and culture which, to an extent, have made man more than the beasts of the jungle. Considered as adaptation, creativity is a failure, a disease and unhappiness; the Vienna historian of culture, Egon Friedell (1927–31) has a brilliant analysis of this point. The maxim of adjustment, equilibrium and homeostasis cannot be followed by anyone who brings one single idea to the earth, including Selye himself, who certainly has paid for doing so.

Life is not a comfortable settling down in pre-ordained grooves of being; at its best, it is *élan vital*, inexorably driven towards a higher form of existence. Admittedly, this is metaphysics and poetic simile; but so, after all, is any image we try to form of the driving forces in the universe.

System-Theoretical Re-orientation

It is along such lines that a new model or image of man seems to be emerging. We may briefly characterize it as the model of man as *active personality system*. This, it appears, is the common denominator of many otherwise different currents such as developmental psychology after Piaget and Werner, various neo-Freudian schools, ego psychology, the "new look" in perception, recent theory of cognition, personality theories such as those of G. Allport and Maslow, new approaches in education, existential psychology and others.

This implies a holistic orientation in psychology. It used to be the general trend of psychology to reduce mental happenings and behavior into a bundle of sensations, drives, innate and learned reactions, or whatever ultimate elements are theoretically presupposed. In contrast, the system concept tries to bring the psychophysiological organism as a whole into the focus of the scientific endeavor.

Thus a new "model of man" appears necessary and, in fact, is slowly emerging in recent trends of humanistic and organismic psychology. Emphasis on the creative side of human beings, on the importance of individual differences, on aspects that are non-utilitarian and beyond the biological values of subsistence and survival—this and more is implied in the model of the active organism. These notions are basic in the re-orientation of psychology which is going on presently; hence the increasing interest general system theory is encountering in psychology and especially psychiatry.

In contrast to the model of the reactive organism expressed by the S-R scheme—behavior as gratification of needs, relaxation of tensions, reestablishment of homeostatic equilibrium, its utilitarian and environmentalistic interpretations, etc.—we come rather to consider the psychophysical organism as a primarily active system. I think human activities cannot be considered otherwise. I, for one, am unable to see how, for example, creative and cultural activities of all sorts can be regarded as "response to stimuli," "gratification of biological needs," "reestablishment of homeostasis" or the like. It does not look particularly "homeostatic" when a businessman follows his restless activities in spite of the ulcers he is developing; or when mankind goes on inventing super-bombs in order to satisfy "biological needs."

The concept applies not only to behavioral, but also to the cognitional aspects. It will be correct to say that it is the general trend in modern psychology and psychiatry, supported by biological insight, to recognize the active part in the cognitive process. Man is not a passive receiver of stimuli coming from an external world, but in a very concrete sense *creates* his universe. This, again, can be expressed in many ways: in Freud's reconstruction of the building-up of the "world" in the child; in terms of developmental psychology according to Piaget, Werner or Schachtel; in terms of the "new look in perception" emphasizing attitudes, affective and motivational factors; in psychology of cognition by analysis of "meaningful learning" after Ausubel; in zoological context by referring to von Uexküll's species-specific *umwelt*; philosophically, in Cassirer's "symbolic forms" and culture-dependent categories; in von Humboldt's and Whorf's evidence of linguistic (i.e. symbolic and cultural) factors in the formation of the experienced universe. "The world as we experience it is the product of perception, not the cause of it." (Cantril, 1962.)

Such a list, in no way complete, illustrates different approaches to throw light on various aspects or facets which eventually should be synthesized. But there is consensus in the general conception. Indeed, if the organism were a camera and cognition a kind of photographic image of the outside world, it would be hard to understand why the cognitive process takes the circuitous route admirably described by Arieti (1965) via fantasmic, mythical and magical universes, only finally and lately to arrive at the supposedly "objective" world outlook of the average American and of Western science.

Such a new "image of man," replacing the robot concept by that of system, emphasizing immanent activity instead of outer-directed reactivity, and recognizing the specificity of human culture compared to animal behavior, should lead to a basic reevaluation of problems of education, training, psychotherapy, and human attitudes in general.

Systems in the Social Sciences

Finally, we should look for the application of the systems conception to the widest perspective, i.e., human groups, societies, and humanity as a whole.

For purposes of discussion, let us understand "social science" in a broad sense, including sociology, economics, political science, social psychology, cultural anthropology, linguistics, a good part of history and the humanities, etc. Let us understand "science" as a nomothetic endeavor, i.e. not a description of singularities but an ordering of facts and elaboration of generalities.

Presupposing these definitions, it may, in my opinion, be stated quite confidently: *Social science is the science of social systems.* For this reason, it will have to use the approach of general systems science.

This appears to be an almost trivial statement, and it can hardly be denied that "contemporary sociological theories" (Sorokin, 1928, 1966) and even their development through history, followed this program. However, proper study of social systems contrasts with two widespread conceptions: first, with atomistic conceptions which neglect study of "relations"; secondly, with conceptions neglecting the specificity of the systems concerned, such as a "social physics" as was often attempted in a reductionist spirit. This requires some comment.

Research into systems of organisms is extensive. It forms an important part of biology, in the study of communities and societies of animals and plants, their growth, competition, struggle for existence, etc., both in the ecological and genetic aspects. Certain aspects of human societies offer themselves for similar considerations; not only aspects so obvious as the growth of human populations but also armament races and warlike conflicts which, according to Richardson and others, can be elaborated in differential equations similar to those used in ecology and, though oversimplified, provide an amount of explanation and even prediction. The spread of rumors can be described by generalized diffusion equations; the flow of automobile traffic can be analyzed in considerations formally corresponding to kinetics and thermodynamics. Such cases are rather typical and straightforward applications of general system theory. However, this is only part of the problem.

Sociology with its allied fields is essentially the study of human groups or systems, from small groups like the family or working crew, over innumerable intermediates of informal and formal organizations to the largest units like nations, power blocks and international relations. The many attempts to provide theoretical formulations are all elaborations of the concept of system or some synonym in this realm. Ultimately the problem of human history looms as the widest possible application of the systems idea.

Concepts and theories provided by the modern systems approach are being increasingly introduced into sociology, such as the concept of general system, of feedback, information, communication, etc.

Present sociological theory largely consists in attempts to define the sociocultural "system," and in discussion of functionalism, i.e., consideration of social phenomena with respect to the "whole" they serve. In the first respect, Sorokin's characterization of sociocultural system as causal-logical-meaningful (as the present author would loosely transcribe it, the biological, the symbolic and value levels) seems best to express the various complexly interconnected aspects.

Functionalist theory has found various expressions as represented by Parsons, Merton, and many others; the recent book by Demerath and Peterson (1968) gives excellent insight into the various currents. The main critique of functionalism, particularly in Parsons' version, is that it overemphasizes maintenance, equilibrium, adjustment, homeostasis, stable institutional structures, and so on, with the result that history, process, sociocultural change, inner-directed development, etc., are underplayed and, at most, appear as "deviants" with a negative value connotation. The theory therefore appears to be one of conservatism and conformism, defending the "system" (or the megamachine of present society, to use Mumford's term) as is, conceptually neglecting and hence obstructing social change. Obviously, general system theory in the form here presented is free of this objection as it incorporates equally maintenance and change, preservation of system and internal conflict; it may therefore be apt to serve as logical skeleton for improved sociological theory (cf. Buckley, 1967).

The practical application, in systems analysis and engineering, of systems theory to problems arising in business, government, international politics, demonstrates that the approach "works" and leads to both understanding and predictions. It especially shows that the systems approach is not limited to material entities in physics, biology and other natural sciences, but is applicable to entities which are partly immaterial and highly heterogeneous. Systems analysis, for example, of a business enterprise encompasses men, machines, buildings, inflow of raw material, outflow of products, monetary values, good will and other imponderables; it may give definite answers and practical advice.

The difficulties are not only in the complexity of phenomena but in the definition of entities under consideration.

At least part of the difficulty is expressed by the fact that the social sciences are concerned with "socio-cultural" systems. Human groups, from the smallest of personal friendships and family to the largest of nations and civilizations, are not only an outcome of social "forces" found, at least in primitive form, in subhuman organisms; they are part of a man-created universe called culture.

Natural science has to do with physical entities in time and space, particles, atoms and molecules, living systems at various levels, as the case may be. Social science has to do with human beings in their self-created universe of culture. The cultural universe is essentially a symbolic universe. Animals are surrounded by a *physical* universe with which they have to cope: physical environment, prey to catch, predators to avoid, and so forth. Man, in contrast, is surrounded by a universe of *symbols*. Starting from language which is the prerequisite of culture, to symbolic relationships with his fellows, social status, laws, science, art, morals, religion and innumerable other things, human behavior, except for the basic aspects of the biological needs of hunger and sex, is governed by symbolic entities.

We may also say that man has values which are more than biological and transcend the sphere of the physical world. These cultural values may be biologically irrelevant or even deleterious: It is hard to see that music, say, has any adaptive or survival value; the values of nation and state become biologically nefarious when they lead to war and to the killing of innumerable human beings.

A System-Theoretical Concept of History

In contrast to biological species which have evolved by way of genetic transformation, only mankind shows the phenomenon of history, which is intimately linked with culture, language and tradition. The reign of nature is dominated by laws progressively revealed in science. Are there laws of history? In view of the fact that laws are relations in a conceptual model or theory, this question is identical with another one: apart from description of happenings, is a *theoretical history* possible? If this is possible at all, it must be an investigation of *systems* as suitable units of research—of human groups, societies, cultures, civilizations, or whatever the appropriate objects of research may be.

A widespread conviction among historians is that this is not so. Science is essentially a *nomothetic* endeavor—it establishes laws based on the fact that events in nature are repeatable and recurrent. In contrast, history does not repeat itself. It has occurred only once, and therefore, history can only be *idiographic*—i.e. a description of events which have occurred in a near or distant past.

Contrary to this opinion, which is the orthodox one of historians, heretics have appeared who held the opposite view and, in one way or another, tried to construct a theoretical history with laws applying to the historical process. This current started with the Italian philosopher Vico in the early 18th century, and continued in the philosophical systems and the investigations by Hegel, Marx, Spengler, Toynbee, Sorokin, Kroeber and others. There are great and obvious differences between these systems. They all agree, however, that the historical process is not completely accidental but follows regularities or laws which can be determined.

As already said, the scientific approach is undisputedly applicable to certain *aspects* of human society. One such field is statistics. We can, and do, formulate many statistical laws or at least regularities for social entities. Population statistics, mortality statistics—without which insurance companies would go bankrupt —Gallup polls, predictions of voting behavior or of the sale of a product show that statistical methods are applicable to a wide range of social phenomena.

Moreover, there are fields where the possibility of a hypothetico-deductive system is generally accepted. One such field is mathematical economics or econometries. The correct system of economics may be disputed, but such systems exist and, as in every science, it is hoped that they will be improved. Mathematical economics also is a case in point of general systems theory which does not concern physical entities. The many-variables problems, different models and mathematical approaches in economics offer a good example of model building and the general systems approach.

Even for those mysterious entities, human values, scientific theories are emerging. In fact, information theory, game theory, and decision theory provide models to deal with aspects of human nd social behavior where the mathematics of classical science s not applicable. Works like Rapoport's *Fights, Games, Debates* (1960) and Boulding's *Conflict and Defence* (1962) present detailed nalyses of phenomena such as armament races, war and war ;ames, competition in the economic and other fields, treated by uch comparatively novel methods.

It is of particular interest that these approaches are concerned vith aspects of human behavior which were believed to be outside)f science: values, rational decisions, information, etc. They are 1ot physicalistic or reductionist. They do not apply physical laws)r use the traditional mathematics of the natural sciences. Rather, 1ew developments of mathematics are emerging, intended to deal ,vith phenomena not encountered in the world of physics.

Again there are uncontested laws with respect to certain im-naterial aspects of culture. For example, language is not a)hysical object, but a product or rather aspect of that intangible :ntity we call human culture. Nevertheless, linguistics tells about laws allowing description, explanation and prediction of ob-served phenomena. Grimm's laws of the consonant mutations in the history of Germanic languages is one of the simpler examples.

In somewhat vaguer form, a lawfulness of cultural events is generally accepted. For example, it appears to be a quite general phenomenon that art goes through a number of stages of archaism, maturity, baroque, and dissolution found in the evolution of art in far-away places and times.

Thus statistical regularities and laws can be found in social phenomena; certain specific aspects can be approached by recent methods, models and techniques which are outside and differ-ent from those of the natural sciences; and we have some ideas about intrinsic, specific and organizational laws of social systems. This is no matter of dispute.

The bone of contention comes in with "theoretical history," the great visions or constructs of history, as those of Vico, Hegel, Marx, Spengler, Toynbee, to mention only some prominent ex-amples. Regularities in "microhistory," i.e., happenings in limited spaces, time-spans, and fields of human activity, certainly are vague, needy of exploration, and far from being exact statements; but their existence is hardly disputable. The attempts at finding regularities in "macrohistory" are almost unequivocally rejected by official history.

Taking away romanticism, metaphysics and moralizing, the

"great systems" appear as models of the historical process, as Toynbee, somewhat belatedly, recognized in the last volume of his *Study*. Conceptual models which, in simplified and therefore comprehensible form, try to represent certain aspects of reality, are basic in any attempt at theory; whether we apply the Newtonian model in mechanics, the model of corpuscle or wave in atomic physics, use simplified models to describe the growth of a population, or the model of a game to describe political decisions. The advantages and dangers of models are well known. The advantage is in the fact that this is the way to create a theory —i.e., the model permits deductions from premises, explanation and prediction, with often unexpected results. The danger is oversimplification: to make it conceptually controllable, we have to reduce reality to a conceptual skeleton—the question remaining whether, in doing so, we have not cut out vital parts of the anatomy. The danger of oversimplification is the greater, the more multifarious and complex the phenomenon is. This applies not only to "grand theories" of culture and history but to models we find in any psychological or sociological journal.

Obviously, the Great Theories are very imperfect models. Factual errors, misinterpretations, fallacies in conclusion are shown in an enormous critical literature and need not concern us here. But taking this criticism for granted, a number of observations still remain.

One thing the various systems of "theoretical history" appear to have demonstrated is the nature of the historical process. History is not a process in an amorphous humanity, or in *Homo sapiens* as a zoological species. Rather it is borne by entities or great systems, called high cultures or civilizations. Their number is uncertain, their delimitations vague, and their interactions complex. But whether Spengler has counted eight great civilizations, Toynbee some 20, Sorokin applies still other categories, or recent research has unveiled so many lost cultures, it appears to be a fact that there was a limited number of cultural entities bearing the historical process, each presenting a sort of life cycle as indeed smaller socio-cultural systems such as businesses, schools in art and even scientific theories certainly do. This course is not a predetermined life span of a thousand years as Spengler maintained (not even individual organisms have a fixed life span but may die earlier or later); nor does it run in splendid isolation.

The extent of cultural diffusion became impressive when archae-ologists explored the prehistoric Amber Road or the Silk Road which dated from the beginning of the Christian era or even earlier, or when they found an Indian Lakshmi statuette in Pompeii and Roman trade stations at the Indian coasts. Expan-sion undreamed of by Spengler and yet by Toynbee, as well as new problems, became apparent in relatively recent years. Certainly the culture of the Khmer, the Etruscans or the pre-Roman Celts deserve their place in the scheme; what was the megalithic culture expanding over the edges of the Mediterranean, the Atlantic and the Baltic Sea, or the Iberian culture which produced as early as 500 B.C. such astonishing work as the Prado's *Lady of Elche?* Nevertheless, there is something like an Egyptian, Greco-Roman, Faustian, Magic, Indian culture (or whatever nomenclature we may prefer), each unique in its "style" (i.e., the unity and whole of its symbolic system), even when absorbing and assimilating culture traits from others and interacting with cultural systems contemporaneous and past.

Furthermore, the ups and downs in history (not exactly cycles or recurrences, but fluctuations) are a matter of public record. As Kroeber (1957) and Sorokin (1950) emphasized, there remains, after subtracting the errors and idiosyncrasies of the philosophers of history, a large area of agreement; and this consists of well-known facts of history. In other words, the disagreements among the theorists of history, and with official history, are not so much a question of data as of interpretation, that is of the models applied. This, after all, is what one would expect according to history of science; for a scientific "revolution," the introduction of a new "paradigm" of scientific thinking (Kuhn, 1962), usually manifests itself in a gamut of competing theories or models.

In such a dispute, the influence of semantics pure and simple should not be underrated. The meaning of the concept of culture itself is a matter of dispute. Kroeber and Kluckhohn (1952) collected and discussed some 160 definitions without coming out with a definitive one. In particular, the anthropologist's and the historian's notions are different. For example, Ruth Benedict's *Patterns of Culture* of New Mexico, British Columbia and Australia are essentially timeless; these patterns have existed since times unrecorded, and if they underwent minor changes in the past, these are outside the scope and methods of the cultural

anthropologist. In contrast, the culture or more properly civilization (to follow idiomatic English) with which the historian is
concerned, is a process in time: the evolution of the Greco-Roman
culture from the Ionian city states to the Roman Empire, of its
plastic art from archaic statuary to Hellenism, of German music
from Bach to Strauss, of science from Copernicus to Einstein, and
so forth. So far as we know, only a limited number of "high
cultures" had and made history—i.e., showed major change in
time, while the hundreds of cultures of the anthropologist remained stationary at their stone and bronze age levels, as the
case may be, before the European impact. In this respect, Spengler
is certainly right, with his concept of culture as a dynamic and
self-evolving entity, against anthropologists to whom one "culture"
—be it of Australian aborigines, of Greece or of the Western
world—is as good as the other, all belonging to one stream of
amorphous humanity with accidental and environment-caused
eddies, rapids and standstills.

Incidentally, such verbal distinctions are more than scholasticism
and have political impact. In Canada, we presently have the
fight about Biculturalism (or of Two Nations, English and
French, in another version). What do we mean? Do we understand culture in the anthropological sense and wish to fight about
tribal differences as they exist between savage peoples in Africa
or Borneo and cause endless warfare and bloodshed? Or do we
mean culture as the French *culture* and German *Kultur*—i.e.,
creative manifestations which still have to be proved, and to be
proved as different between English- and French-Canadians?
Clearly, political opinions and decisions will largely depend on
the definition. The concept of nation in the U.N. has been based
on the "anthropological" notion (if not on arbitrary frontiers left
from the Colonial period); the result has been somewhat less
than encouraging.

Another semantic problem is implied in "organismic" theories
of sociology and history. Spengler called the great civilizations
organisms with a life cycle including birth, growth, maturity,
senescence and death; an enormous host of critics proved the
obvious, namely, that cultures are not organisms like animals
or plants, individual entities well-bounded in time and space.
In contrast, the organismic conception is rather well-treated in
sociology because its metaphorical character is understood. A
business firm or manufacturing plant is a "system" and therefore

shows "organismic" features; but the difference of a "plant" in the botanist's and industrialist's sense is too obvious to present a problem. The fight would hardly have been possible in the French language where it is good usage to speak of the *organisme* of an institution (say the postal service), of a commercial firm or a professional association; the metaphor is understood and, therefore, not a subject of dispute.

Instead of emphasizing the shortcomings of the cyclic historians which are rather natural in an embryonic stage of the science, it seems more profitable to emphasize their agreement in many respects. One point of agreement makes the issue into more than an academic question. This issue has, as it were, touched a raw nerve, and gained for Toynbee and Spengler both popular acclaim and an emotional reaction otherwise uncommon in academic debate. It is the thesis expressed in Spengler's title, *The Decline of the West*, the statement that in spite or perhaps because of our magnificent technological achievements, we live in a time of cultural decay and impending catastrophe.

The Future in System-Theoretical Aspect

The dominance of mass man and the suppression of the individual by an ever expanding social machinery; the breakdown of the traditional system of values and its replacement by pseudo-religions, ranging from nationalism to the cult of status symbols, to astrology, psychoanalysis and Californian sectarianism; the decay of creativity in art, music and poetry; the willing submission of the mass to authoritarianism, be it a dictator or an impersonal élite; the colossal fights between a decreasing number of super-states: these are some of the symptoms recurring in our days. "We note the psychological change in those classes of society which had been up till then the creators of culture. Their creative power and creative energy dry up; men grow weary and lose interest in creation and cease to value it; they are disenchanted; their effort is no longer an effort toward a creative ideal for the benefit of humanity, their minds are occupied either with material interests, or with ideals unconnected with life on earth and realized elsewhere." This is not an editorial of yesterday's newspaper, but a description of the decay of the Roman Empire by its well-known historian, Rostovtzeff.

However, against these and other symptoms catalogued by the

prophets of doom, there are two factors in which our civilization obviously is unique in comparison to those that have perished in the past. One is the *technological development* which permits a control of nature never before achieved and would open a way to alleviate the hunger, disease, overpopulation, etc. to which humanity was previously exposed. The other factor is the *global nature* of our civilization. Previous ones were limited by geographical boundaries, and comprised only limited groups of human beings. Our civilization comprises the whole planet and even reaches beyond in the conquest of space. Our technological civilization is not the privilege of comparatively small groups such as the citizens of Athens or of the Roman Empire, of Germans or French, or of white Europeans. Rather it is open to all human beings of whatever color, race or creed.

These are indeed singularities which explode the cyclic scheme of history and seem to place our civilization at a different level from previous ones. Let us try an admittedly tentative synthesis.

I believe the "decline of the west" is not a hypothesis or a prophesy—it is an accomplished fact. That splendid cultural development which started in the European countries around the year 1000 and produced Gothic cathedrals, Renaissance art, Shakespeare and Goethe, the precise architecture of Newtonian physics and all the glory of European culture—this enormous cycle of history is accomplished and cannot be revivified by artificial means.

We have to reckon with the stark reality of a mass civilization, technological, international, encompassing the earth and all of mankind, in which cultural values and creativity of old are replaced by novel devices. The present power struggles may, in their present explosive phase, lead to universal atomic devastation. If not, the differences between West and East probably will, one way or the other, become insignificant because the similarity of material culture in the long run will prove stronger than ideological differences.

9 General System Theory in Psychology and Psychiatry

The Quandary of Modern Psychology

In recent years the concept of "system" has gained increasing influence in psychology and psychopathology. Numerous investigations have referred to general system theory or to some part of it (for example, F. Allport, 1955; G. W. Allport, 1960; Anderson, 1957; Arieti, 1962; Brunswik, 1956; Bühler, 1959; Krech, 1950; Lennard & Bernstein, 1960; Menninger, 1957; Menninger et al., 1958; Miller, 1955; Pumpian-Mindlin, 1959; Syz, 1963). Gordon W. Allport ended the reedition of his classic (1961) with "Personality as System"; Karl Menninger (1963) based his system of psychiatry on general system theory and organismic biology; Rapaport (1960) even spoke of the "epidemiclike popularity in psychology of open systems" (p. 144). The question arises why such a trend has appeared.

American psychology in the first half of the 20th century was dominated by the concept of the reactive organism, or, more dramatically, by the model of man as a robot. This conception was common to all major schools of American psychology, classical and neobehaviorism, learning and motivation theories, psychoanalysis, cybernetics, the concept of the brain as a computer, and so forth. According to a leading personality theorist,

Man is a computer, an animal, or an infant. His destiny is completely determined by genes, instincts, accidents, early conditionings and reinforcements, cultural and social forces. Love is a secondary drive based on hunger and oral sensations or a reaction formation to an innate underlying hate. In the majority of our personological formulations there are no provisions for creativity, no admitted margins of freedom for voluntary decisions, no fitting recognitions of the power of ideals, no bases for selfless actions, no ground at all for any hope that the human race can save itself from the fatality that now confronts it. If we psychologists were all the time, consciously or unconsciously, intending out of malice to reduce the concept of human nature to its lowest common denominator, and were gloating over our success in so doing, then we might have to admit that to this extent the Satanic spirit was alive within us. (Murray, 1962, pp. 36–54)

The tenets of robot psychology have been extensively criticized; for a survey of the argument, the reader may consult Allport's well-balanced evaluations (1955, 1957, 1961) and the recent historical outline by Matson (1964) which is both brilliantly written and well documented. The theory nevertheless remained dominant for obvious reasons. The concept of man as robot was both an expression of and a powerful motive force in industrialized mass society. It was the basis for behavioral engineering in commercial, economic, political, and other advertising and propaganda; the expanding economy of the "affluent society" could not subsist without such manipulation. Only by manipulating humans ever more into Skinnerian rats, robots, buying automata, homeostatically adjusted conformers and opportunists (or, bluntly speaking, into morons and zombies) can this great society follow its progress toward ever increasing gross national product. As a matter of fact (Henry, 1963), the principles of academic psychology were identical with those of the "pecuniary conception of man" (p. 45ff.).

Modern society provided a large-scale experiment in manipulative psychology. If its principles are correct, conditions of tension and stress should lead to increase of mental disorder. On the other hand, mental health should be improved when basic needs for food, shelter, personal security, and so forth, are satisfied; when repression of infantile instincts is avoided by permissive training in bodily functions; when scholastic demands are reduced so as not to overload a tender mind; when sexual gratification is provided at an early age, and so on.

The behavioristic experiment led to results contrary to expectation. World War II—a period of extreme physiological and psychological stress—did not produce an increase in neurotic (Opler, 1956) or psychotic (Llavero, 1957) disorders, apart from direct shock effects such as combat neuroses. In contrast, the affluent society produced an unprecedented number of mentally ill. Precisely under conditions of reduction of tensions and gratification of biological needs, novel forms of mental disorder appeared as existential neurosis, malignant boredom, and retirement neurosis (Alexander, 1960), i.e., forms of mental dysfunction originating not from repressed drives, from unfulfilled needs, or from stress but from the meaninglessness of life. There is the suspicion (Arieti, 1959, p. 474; von Bertalanffy, 1960a) (although not substantiated statistically) that the recent increase in schizophrenia may be caused by the "other-directedness" of man in modern society. And there is no doubt that in the field of character disorders, a new type of juvenile delinquency has appeared: crime not for want or passion, but for the fun of it, for "getting a kick," and born from the emptiness of life (Anonymous, *Crime and Criminologists*, 1963; Hacker, 1955).

Thus theoretical as well as applied psychology was led into malaise regarding basic principles. This discomfort and the trend toward a new orientation were expressed in many different ways such as in the various neo-Freudian schools, ego psychology, personality theories (Murray, Allport), the belated reception of European developmental and child psychology (Piaget, Werner, Charlotte Bühler), the "new look" in perception, self-realization (Goldstein, Maslow), client-centered therapy (Rogers), phenomenological and existential approaches, sociological concepts of man (Sorokin, 1963), and others. In the variety of modern currents, there is one common principle: to take man not as reactive automaton or robot but as *an active personality system.*

The reason for the current interest in general system theory therefore appears to be that it is hoped that it may contribute toward a more adequate conceptual framework for normal and pathological psychology.

System Concepts in Psychopathology

General system theory has its roots in the organismic conception in biology. On the European continent, this was developed by the present author (1928a) in the 1920's, with parallel developments in the Anglo-Saxon countries (Whitehead, Woodger, Coghill and others) and in psychological gestalt theory (W. Köhler). It is interesting to note that Eugen Bleuler (1931) followed with sympathetic interest this development in its early phase. A similar development in psychiatry was represented by Goldstein (1939).

ORGANISM AND PERSONALITY

In contrast to physical forces like gravity or electricity, the phenomena of life are found only in individual entities called organisms. Any organism is a system, that is, a dynamic order of parts and processes standing in mutual interaction (Bertalanffy, 1949a, p. 11). Similarly, psychological phenomena are found only in individualized entities which in man are called personalities. "Whatever else personality may be, it has the properties of a system" (G. Allport, 1961, p. 109).

The "molar" concept of the psychophysical organism as system contrasts with its conception as a mere aggregate of "molecular" units such as reflexes, sensations, brain centers, drives, reinforced responses, traits, factors, and the like. Psychopathology clearly shows mental dysfunction as a system disturbance rather than as a loss of single functions. Even in localized traumas (for example, cortical lesions), the ensuing effect is impairment of the total action system, particularly with respect to higher and, hence, more demanding functions. Conversely, the system has considerable regulative capacities (Bethe, 1931; Goldstein, 1959; Lashley, 1929).

THE ACTIVE ORGANISM

"Even without external stimuli, the organism is not a passive but an intrinsically active system. Reflex theory has presupposed that the primary element of behavior is response to external stimuli. In contrast, recent research shows with increasing clarity that autonomous activity of the nervous system, resting in the system itself, is to be considered primary. In evolution and development, reactive mechanisms appear to be superimposed upon primitive, rhythmic-locomotor activities. The stimulus (i.e., a change in external conditions) does not *cause* a process in an otherwise inert system; it only *modifies* processes in an autonomously active system" (Bertalanffy, 1937, pp. 133 ff.; also 1960).

The living organism maintains a disequilibrium called the steady state of an open system and thus is able to dispense existing potentials or "tensions" in spontaneous activity or in response to releasing stimuli; it even advances toward higher order and organization. The robot model considers response to stimuli, reduction of tensions, reestablishment of an equilibrium disturbed by outside factors, adjustment to environment, and the like, as the basic and universal scheme of behavior. The robot model, however, only partly covers animal behavior and does not cover an essential portion of human behavior at all. The insight into the primary immanent activity of the psychophysical organism necessitates a basic reorientation which can be supported by any amount of biological, neurophysiological, behavioral, psychological, and psychiatric evidence.

Autonomous activity is the most primitive form of behavior (von Bertalanffy, 1949a; Carmichael, 1954; Herrick, 1956; von Holst, 1937; Schiller, 1957; H. Werner 1957a); it is found in brain function (Hebb, 1949) and in psychological processes. The discovery of activating systems in the brain stem (Berlyne, 1960; Hebb, 1955; Magoun, 1958) has emphasized this fact in recent years. Natural behavior encompasses innumerable activities beyond the S-R scheme, from exploring, play, and rituals in animals (Schiller, 1957) to economic, intellectual, esthetic, religious, and the like pursuits to self-realization and creativity in man. Even rats seem to "look" for problems (Hebb, 1955), and the healthy child and adult are going far beyond the reduction of tensions or gratification of needs in innumerable activities that cannot be reduced to primary or secondary drives (G. Allport, 1961, p. 90). All such behavior is performed for its own sake, deriving gratification ("function pleasure," after K. Bühler) from the performance itself.

For similar reasons, complete relaxation of tensions as in sensory-deprivation experiments is not an ideal state but is apt to produce insufferable anxiety, hallucinations, and other psychosislike symptoms. Prisoner's psychosis, or exacerbation of symptoms in the closed ward, and retirement and weekend neurosis are related clinical conditions attesting that the psychophysical organism needs an amount of tension and activity for healthy existence.

It is a symptom of mental disease that spontaneity is impaired. The patient increasingly *becomes* an automaton or S-R machine, is pushed by biological drives, obsessed by needs for food, elimination, sex gratification, and so on. The model of the passive organism is a quite adequate description of the stereotype behavior of compulsives, of patients with brain lesions, and of the waning of autonomous activity in catatonia and related psychopathology. But by the same token, this emphasizes that normal behavior is different.

HOMEOSTASIS

Many psychophysiological regulations follow the principles of homeostasis. However, there are apparent limitations (cf. pp. 160ff.). Generally, the homeostasis scheme is not applicable (1) to dynamic regulations—i.e., regulations not based upon fixed mechanisms but taking place within a system functioning as a whole (for example, regulative processes after brain lesions); (2) to spontaneous activities; (3) to processes whose goal is not reduction but is building up of tensions; and (4) to processes of growth, development, creation, and the like. We may also say that homeostasis is inappropriate as an explanatory principle for those human activities which are nonutilitarian—i.e., not serving the primary needs of self-preservation and survival and their secondary derivatives, as is the case with many cultural manifestations. The evolution of Greek sculpture, Renaissance painting, or German music had nothing to do with adjustment or survival because they are of symbolic rather than biological value (Bertalanffy, 1959; also 1964c) (compare below). But even living nature is by no means merely utilitarian (von Bertalanffy, 1949a, pp. 106ff).

The principle of homeostasis has sometimes been inflated to a point where it becomes silly. The martyr's death at the stake is explained (Freeman, 1948) "by abnormal displacement" of his internal processes so that death is more "homeostating" than continuing existence (pp. 142ff.); the mountain climber is supposed to risk his life because "losing valued social status may be more upsetting" (Stagner, 1951). Such examples show to what extremes some writers are willing to go in order to save a scheme which is rooted in economic-commercial philosophy and sets a premium on conformity and opportunism as ultimate values. It should not be forgotten that Cannon (1932), eminent physiologist and thinker that he was, is free of such distortions; he explicitly emphasized the "priceless unessentials" beyond homeostasis (p. 323) (cf. also Frankl, 1959b; Toch and Hastorf, 1955).

The homeostasis model is applicable in psychopathology because nonhomeostatic functions, as a rule, decline in mental patients. Thus Karl Menninger (1963) was able to describe the progress of mental disease as a series of defense mechanisms, settling down at ever lower homeostatic levels until mere preservation of physiological life is left. Arieti's (1959) concept of progressive teleological regression in schizophrenia is similar.

DIFFERENTIATION

"Differentiation is transformation from a more general and homeogeneous to a more special and heterogeneous condition" (Conklin after Cowdry, 1955, p. 12). "Wherever development occurs it proceeds from a state of relative globality and lack of differentiation to a state of increasing differentiation, articulation, and hierarchic order" (H. Werner, 1957b).

The principle of differentiation is ubiquitous in biology, the evolution and development of the nervous system, behavior, psychology, and culture. We owe to Werner (1957a) the insight that mental functions generally progress from a syncretic state where percepts, motivation, feeling, imagery, symbols, concepts, and so forth are an amorphous unity, toward an ever clearer distinction of these functions. In perception the primitive state seems to be one of synesthesia (traces of which are left in the human adult and which may reappear in schizophrenia, mescaline, and LSD experience) out of which visual, auditional, tactual, chemical, and other experiences are separated.* In animal and a good deal of human behavior, there is a perceptual-

*Cf. recently J. J. Gibson, *The Senses Considered as Perceptual Systems*, (Boston, Houghton Mifflin, 1966) ; the model of neural hologram in brain physiology (K. H. Pribram, "Four R's of Remembering" in *The Neurophysiological and Biochemical Bases of Learning*, Cambridge, Harvard University Press) , and so on.

emotive-motivational unity; perceived objects without emotional-motivational undertones are a late achievement of mature, civilized man. The origins of language are obscure; but insofar as we can form an idea, it seems that "holophrastic" (W. Humboldt, cf. Werner, 1957a) language and thought—i.e., utterances and thoughts with a broad aura of associations—preceded separation of meanings and articulate speech. Similarly, the categories of developed mental life such as the distinction of "I" and objects, space, time, number, causality, and so forth, evolved from a perceptual-conceptual-motivational continuum represented by the "paleologic" perception of infants, primitives, and schizophrenics (Arieti, 1959; Piaget, 1959; Werner, 1957a). Myth was the prolific chaos from which language, magic, art, science, medicine, mores, morals, and religion were differentiated (Cassirer, 1953–1957).

Thus "I" and "the world," "mind" and "matter," or Descartes's *"res cogitans"* and *"res extensa"* are not a simple datum and primordial antithesis. They are the final outcome of a long process in biological evolution, mental development of the child, and cultural and linguistic history, wherein the perceiver is not simply a receptor of stimuli but in a very real sense *creates* his world (for example Bruner, 1958; Cantril, 1962; Geertz, 1962; Matson, 1964, pp. 181ff.). The story can be told in different ways (for example, G. Allport, 1961, pp. 110–138; von Bertalanffy, 1964a and 1965; Cassirer, 1953–1957; Freud, 1920; Merloo, 1956, pp. 196–199; Piaget, 1959; Werner, 1957a), but there is general agreement that differentiation arose from an "undifferentiated absolute of self and environment" (Berlyne, 1957), and that the animistic experience of the child and the primitive (persisting still in Aristotelian philosophy), the "physiognomic" outlook (Werner, 1957a), the experience of "we" and "thou" (still much stronger in Oriental than in Western thinking—Koestler, 1960), empathy, etc., were steps on the way until Renaissance physics eventually "discovered inanimate nature." "Things" and "self" emerge by a slow build-up of innumerable factors of gestalt dynamics, of learning processes, and of social, cultural, and linguistic determinants; the full distinction between "public objects" and "private self" is certainly not achieved without naming and language, that is, processes at the symbolic level; and perhaps this distinction presupposes a language of the Indo-Germanic type (Whorf, 1956).

In psychopathology and schizophrenia, all these primitive states may reappear by way of regression and in bizarre manifestations; bizarre because there are arbitrary combinations of archaic elements among themselves and with more sophisticated thought processes. On the other hand, the experience of the child, savage, and non-Westerner, though primitive, nevertheless forms an organized universe. This leads to the next group of concepts to be considered.

CENTRALIZATION AND RELATED CONCEPTS

"Organisms *are* not machines; but they can to a certain extent *become* machines, congeal into machines. Never completely, however; for a thoroughly mechanized organism would be incapable of reacting to the incessantly changing conditions of the outside world" (von Bertalanffy, 1949a, pp. 17ff.). The *principle of progressive mechanization* expresses the transition from undifferentiated wholeness to higher function, made possible by specialization and "division of labor"; this principle implies also loss of potentialities in the components and of regulability in the whole.

Mechanization frequently leads to establishment of *leading parts*, that is, components dominating the behavior of the system. Such centers may exert "trigger causality," i.e., in contradistinction to the principle, *causa aequat effectum*, a small change in a leading part may by way of *amplification mechanisms* cause large changes in the total system. In this way, a *hierarchic order* of parts or processes may be established (cf. Chapter 3). These concepts hardly need comment except for the one debated point.

In the brain as well as in mental function, centralization and hierarchic order are achieved by stratification (A. Gilbert, 1957; Lersch, 1960; Luthe, 1957; Rothacker, 1947), i.e., by superimposition of higher "layers" that take the role of leading parts. Particulars and disputed points are beyond the present survey. However, one will agree that—in gross oversimplification—three major layers, or evolutionary steps, can be distinguished. In the brain these are (1) the paleencephalon, in lower vertebrates, (2) the neencephalon (cortex), evolving from reptiles to mammals, and (3) certain "highest" centers, especially the motoric speech (Broca's) region and the large association areas which are found only in man. Concurrently there is an anterior shift of controlling centers, for example, in the apparatus of vision from the colliculi optici of the mesencephalon (lower vertebrates) to the corpora geniculata lateralia of the diencephalon (mammals) to the regio calcarina of the telencephalon (man).*

In some way parallel is stratification in the mental system which can be roughly circumscribed as the domains of instincts, drives, emotions, the primeval "depth personality"; perception and voluntary action; and the symbolic activities characteristic of man. None of the available formulations (for example, Freud's id, ego, and superego, and those of German stratification theorists) is unobjectionable. The neurophysiological meaning of a small portion of brain processes being "conscious" is completely unknown. The Freudian unconscious, or id, comprises only limited aspects and already pre-Freudian authors have given a much more comprehensive survey of unconscious functions (Whyte, 1960). Although these problems need further clarification, it is incorrect when Anglo-Saxon authors refuse stratification for being "philosophical" (Eysenck, 1957) or insist that there is no fundamental difference between the behavior of rat and that of man (Skinner, 1963). Such an attitude simply ignores elementary zoological facts. Moreover, stratification is indispensable for understanding psychiatric disturbances.

REGRESSION

The psychotic state is sometimes said to be a "regression to older and more infantile forms of behavior." This is incorrect; already E. Bleuler noted that the child is not a little schizophrenic but a normally functioning though primitive being. "The schizophrenic will regress to, but not integrate at, a lower level; he will remain disorganized" (Arieti, 1959, p. 475). Regression is essentially disintegration of personality; that is, *dedifferentiation* and *decentralization*. Dedifferentiation means that there is not a loss of meristic functions, but a reappearance of primitive states (syncretism, synesthesia, paleologic thinking, and so forth). Decentralization is, in the extreme, functional dysencephalization

*Cf. recently A. Koestler, *The Ghost in the Machine*, (London, Hutchinson, 1967).

in the schizophrenic (Arieti, 1955). Splitting of personality, according to E. Bleuler, in milder form neurotic complexes (i.e., psychological entities that assume dominance), disturbed ego function, weak ego, and so forth, similarly indicate loosening of the hierarchic mental organization.

BOUNDARIES

Any system as an entity which can be investigated in its own right must have boundaries, either spatial or dynamic. Strictly speaking, spatial boundaries exist only in naive observation, and all boundaries are ultimately dynamic. One cannot exactly draw the boundaries of an atom (with valences sticking out, as it were, to attract other atoms), of a stone (an aggregate of molecules and atoms which mostly consist of empty space, with particles in planetary distances), or of an organism (continually exchanging matter with environment).

In psychology, the boundary of the ego is both fundamental and precarious. As already noted, it is slowly established in evolution and development and is never completely fixed. It originates in proprioceptive experience and in the body image, but self-identity is not completely established before the "I," "Thou," and "it" are named. Psychopathology shows the paradox that the ego boundary is at once too fluid and too rigid. Syncretic perception, animistic feeling, delusions and hallucinations, and so on, make for insecurity of the ego boundary; but within his self-created universe the schizophrenic lives "in a shell," much in the way animals live in the "soap bubbles" of their organization-bound worlds (Schiller, 1957). In contrast to the animal's limited "ambient," man is "open to the world" or has a "universe"; that is, his world widely transcends biological bondage and even the limitations of his senses. To him, "encapsulation" (Royce, 1964) —from the specialist to the neurotic, and in the extreme, to the schizophrenic—sometimes is a pathogenic limitation of potentialities. These are based in man's symbolic functions.

SYMBOLIC ACTIVITIES

"Except for the immediate satisfaction of biological needs, man lives in a world not of things but of symbols" (von Bertalanffy,

1956a). We may also say that the various symbolic universes, material and non-material, which distinguish human cultures from animal societies, are part, and easily the most important part, of man's behavior system. It can be justly questioned whether man is a rational animal; but he certainly is a symbol-creating and symbol-dominated being throughout.

Symbolism is recognized as the unique criterion of man by biologists (von Bertalanffy, 1956a; Herrick, 1956), physiologists of the Pavlovian school ("secondary signal system") (Luria, 1961), psychiatrists (Appleby, Scher & Cummings, 1960; Arieti, 1959; Goldstein, 1959), and philosophers (Cassirer, 1953–1957; Langer, 1942). It is not found even in leading textbooks of psychology in consequence of the predominant robot philosophy. But it is precisely for symbolic functions that "motives in animals will not be an adequate model for motives in man" (G. Allport, 1961, p. 221), and that human personality is not finished at the age of three or so, as Freud's instinct theory assumed.

The definition of symbolic activities will not be discussed here; the author has attempted to do so elsewhere (von Bertalanffy, 1956a and 1965). It suffices to say that probably all notions used to characterize human behavior are consequences or different aspects of symbolic activity. Culture or civilization; creative proception in contrast to passive perception (Murray, G. W. Allport), objectivation of both things outside and the self (Thumb, 1943), ego-world unity (Nuttin, 1957), abstract against concrete stratum (Goldstein, 1959); having a past and future, "time-binding," anticipation of future; true (Aristotelian) purposiveness (cf. Chapter 3), intention as conscious planning (G. Allport, 1961, p. 224); dread of death, suicide; will to meaning (Frankl, 1959b), interest as engaging in self-gratifying cultural activity (G. Allport, 1961, p. 225), idealistic devotion to a (perhaps hopeless) cause, martyrdom; "forward trust of mature motivation" (G. Allport, 1961, p. 90); self-transcendence; ego autonomy, conflict-free ego functions; essential aggression (von Bertalanffy, 1958); conscience, superego, ego ideal, values, morals, dissimulation, truth and lying—all these stem from the root of creative symbolic universes and can therefore not be reduced to biological drives, psychoanalytic instincts, reinforcement of gratifications, or other biological factors. The distinction of *biological* and *specific human values* is that the former concerns the maintenance of the individual and the survival of the species; the latter always concerns a symbolic universe (Bertalanffy, 1959 and 1964c).

In consequence, mental disturbances in man, as a rule, involve disturbances of symbolic functions. Kubie (1953), appears to be correct when, as a "new hypothesis" on neuroses, he distinguishes "psychopathological processes which arise through the distorting impact of highly charged experiences at an early age" from those "consisting in the distortion of symbolic functions." Disturbances in schizophrenia are essentially also at the symbolic level and able to take many different forms: Loosening of associational structure, breakdown of the ego boundary, speech and thought disturbances, concretization of ideas, desymbolization, paleologic thinking, and others. We refer to Arieti's (1959) and Goldstein's (1959) discussions.

The conclusion (which is by no means generally accepted) is that mental illness is a *specifically human phenomenon*. Animals may behaviorally show (and for all we know by empathy experience) any number of perceptional, motoric and mood disturbances, hallucinations, dreams, faulty reactions, and the like. Animals cannot have the disturbances of symbolic functions that are essential ingredients of mental disease. In animals there cannot be disturbance of ideas, delusions of grandeur or of persecution, etc., for the simple reason that there are no ideas to start with. Hence, "animal neurosis" is only a partial model of the clinical entity (von Bertalanffy, 1957a).

This is the ultimate reason why human behavior and psychology cannot be reduced to biologistic notions like restoration of homeostasis, conflict of biological drives, unsatisfactory mother-infant relationships, and the like. Another consequence is the culture-dependence of mental illness both in symptomatology and epidemiology. To say that psychiatry has a physio-psycho-sociological framework is but another expression of the same fact.

For the same reason, human striving is more than self-realization; it is directed toward objective goals and realization of values (Frankl, 1959a, 1959b; 1960), which means nothing else than symbolic entities which in a way become detached from their creators (von Bertalanffy, 1956a; also 1965). Perhaps we may venture a definition. There may be conflict between biological drives and a symbolic value system; this is the situation of psychoneurosis. Or there may be conflict between symbolic universes, or loss of value orientation and experience of meaninglessness in the individual; this is the situation when existential or "noogenic" neurosis arises. Similar considerations apply to "character disorders" like juvenile delinquency that, quite apart from their psychodynamics, stem from the breakdown or erosion of the value system. Among other things, culture is an important psycho-hygienic factor (von Bertalanffy, 1959 and 1964c).

System—A New Conceptual Framework

Having gone through a primer of system-theoretical notions, we may summarize that these appear to provide a consistent framework for psychopathology.

Mental disease is essentially a disturbance of system functions of the psychophysical organism. For this reason, isolated symptoms or syndromes do not define the disease entity (von Bertalanffy, 1960a). Look at some classical symptoms of schizophrenia. "Loosening of associational structure" (E. Bleuler) and unbridled chains of associations; quite similar examples are found in "purple" poetry and rhetoric. Auditory hallucinations; "voices" told Joan of Arc to liberate France. Piercing sensations; a great mystic like St. Teresa reported identical experience. Fantastic world constructions; those of science surpass any schizophrenic's. This is not to play on the theme "genius and madness," but it is apt to show that not single criteria but integration makes for the difference.

Psychiatric disturbances can be neatly defined in terms of system functions. In reference to *cognition*, the worlds of psychotics, as impressively described by writers of the phenomenological and existentialist schools (for example, May *et al.*, 1958), are "products of their brains." But our normal world is shaped also by emotional, motivational, social, cultural, linguistic, and the like factors, amalgamated with perception proper. Illusions and delusions, and hallucinations at least in dreams, are present in the healthy individual; the mechanisms of illusion play even an important role in constancy phenomena, without which a consistent world image would be impossible. The contrast of normality to schizophrenia is not that normal perception is a plane mirror of reality "as is," but that schizophrenia has subjective elements that run wild and that are disintegrated.

The same applies at the symbolic level. Scientific notions such as the earth running with unimaginable speed through the universe or a solid body consisting mostly of empty space interlaced with tiny energy specks at astronomical distances, contradict all everyday experience and "common sense" and are more fantastic than the "world designs" of schizophrenics. Nevertheless the scientific notions happen to be "true"—i.e., they fit into an integrated scheme.

Similar considerations apply to *motivation*. The concept of spontaneity draws the borderline. Normal motivation implies autonomous activity, integration of behavior, plasticity in and adaptability to changing situations, free use of symbolic anticipation, decision, and so forth. This emphasizes the hierarchy of functions, especially the symbolic level superimposed upon the organismic. Hence beside the organismic principle of "spontaneous activity" the "humanistic" principle of "symbolic functions" must be basic in system-theoretical consideration.

Hence the answer whether an individual is mentally sound or not is ultimately determined by whether he has *an integrated universe consistent within the given cultural framework* (von Bertalanffy, 1960a). So far as we can see, this criterion comprises all phenomena of psychopathology as compared with normality and leaves room for culture-dependence of mental norms. What may be consistent in one culture may be pathological in another, as cultural anthropologists (Benedict, 1934) have shown.

This concept has definite implications for *psychotherapy*. If the psychophysical organism is an active system, occupational and adjunctive therapies are an obvious consequence; evocation of creative potentialities will be more important than passive adjustment. If these concepts are correct, more important than "digging the past" will be insight into present conflicts, attempts at reintegration, and orientation toward goals and the future, that is, symbolic anticipation. This, of course, is a paraphrase of recent trends in psychotherapy which thus may be grounded in "personality as system." If, finally, much of present neurosis is "existential," resulting from meaninglessness of life, then "logotherapy" (Frankl, 1959b), i.e., therapy at the symbolic level, will be in place.

It therefore appears that—without falling into the trap of

"nothing-but" philosophy and disparaging other conceptions—a system theory of personality provides a sound basis for psychology and psychopathology.

Conclusion

System theory in psychology and psychiatry is not a dramatic dénouement or new discovery, and if the reader has a *déjà vu* feeling, we shall not contradict him. It was our intention to show that system concepts in this field are not speculation, are not an attempt to press facts into the straitjacket of a theory which happens to be in vogue, and have nothing to do with "mentalistic anthropomorphism," so feared by behaviorists. Nevertheless, the system concept is a radical reversal with respect to robotic theories, leading to a more realistic (and incidentally more dignified) image of man. Moreover, it entails far-reaching consequences for the scientific world view which can only be alluded to in the present outline:

(1) The system concept provides a theoretical framework which is *psychophysically neutral*. Physical and physiological terms such action potentials, chemical transmission at synapses, neural network, and the like are not applicable to mental phenomena, and even less can psychological notions be applied to physical phenomena. System terms and principles like those discussed can be applied to facts in either field.

(2) The mind-body problem cannot be discussed here, and the author has to refer to another investigation (von Bertalanffy, 1964a). We can only summarize that the *Cartesian dualism* between matter and mind, objects outside and ego inside, brain and consciousness, and so forth, is incorrect both in the light of direct phenomenological experience and of modern research in various fields; it is a conceptualization stemming from 17th-century physics which, even though still prevailing in modern debates (Hook, 1961; Scher, 1962), is obsolete. In the modern view, science does not make metaphysical statements, whether of the materialistic, idealistic, or positivistic sense-data variety. It is a conceptual construct to reproduce limited aspects of experience in their formal structure. Theories of behavior and of psychology should be similar in their formal structure or isomorphic. Possibly systems concepts are the first beginning of such "common

language" (compare Piaget and Bertalanffy in Tanner and Inhelder, 1960). In the remote future this may lead to a "unified theory" (Whyte, 1960) from which eventually material and mental, conscious and unconscious aspects could be derived.

(3) Within the framework developed, the problem of *free will* or *determinism* also receives a new and definite meaning. It is a pseudo-problem, resulting from confusion of different levels of experience and of epistemology and metaphysics. We *experience* ourselves as free, for the simple reason that the category of causality is not applied in direct or immediate experience. Causality is a category applied to bring order into objectivated experience reproduced in symbols. Within the latter, we try to *explain* mental and behavioral phenomena as causally determined and can do so with increasing approximation by taking into account ever more factors of motivation, by refining conceptual models, etc. Will is not *determined*, but is *determinable*, particularly in the machine-like and average aspects of behavior, as motivation researchers and statisticians know. However, causality is not metaphysical necessity, but is one instrument to bring order into experience, and there are other "perspectives" (Chapter 10), of equal or superior standing.

(4) Separate from the epistemological question is the moral and legal question of *responsibility*. Responsibility is always judged within a symbolic framework of values as accepted in a society under given circumstances. For example, the M'Naghten rules which excuse the offender if "he cannot tell right from wrong," actually mean that the criminal goes unpunished if his symbolic comprehension is obliterated; hence his behavior is determined only by "animal" drives. Killing is prohibited and is punished as murder within the symbolic framework of the ordinary state of society, but is commanded (and refusal of the command is punished) in the different value frame of war.

10 The Relativity of Categories

The Whorfian Hypothesis

Among recent developments in the anthropological sciences, hardly any have found so much attention and led to so much controversy as have the views advanced by the late Benjamin Whorf.

The hypothesis offered by Whorf is,

> that the commonly held belief that the cognitive processes of all human beings possess a common logical structure which operates prior to and independently of communication through language, is erroneous. It is Whorf's view that the linguistic patterns themselves determine what the individual perceives in this world and how he thinks about it. Since these patterns vary widely, the modes of thinking and perceiving in groups utilizing different linguistic systems will result in basically different world views (Fearing, 1954).

> We are thus introduced to a new principle of relativity which holds that all observers are not led by the same physical evidence to the same picture of the universe, unless their linguistic backgrounds are similar. . . . We cut up and organize the spread and flow of events as we do largely because, through

our mother tongue, we are parties of an agreement to do so, not because nature itself is segmented in exactly that way for all to see (Whorf, 1952, p. 21).

For example, in the Indo-European languages substantives, adjectives and verbs appear as basic grammatic units, a sentence being essentially a combination of these parts. This scheme of a persisting entity separable from its properties and active or passive behavior is fundamental for the categories of occidental thinking, from Aristotle's categories of "substance," "attributes" and "action" to the antithesis of matter and force, mass and energy in physics.

Indian languages, such at Nootka (Vancouver Island) or Hopi, do not have parts of speech or separable subject and predicate. Rather they signify an event as a whole. When we say, "a light flashed" or "it (a dubious hypostatized entity) flashed," Hopi uses a single term, "flash (occurred) ."[1]

It would be important to apply the methods of mathematical logic to such languages. Can statements in languages like Nootka or Hopi be rendered by the usual logistic notation, or is the latter itself a formalization of the structure of Indo-European language? It appears that this important subject has not been investigated.

Indo-European languages emphasize time. The "give-and-take" between language and culture leads, according to Whorf, to keeping of records, diaries, mathematics stimulated by accounting; to calendars, clocks, chronology, time as used in physics; to the historical attitude, interest in the past, archeology, etc. It is interesting to compare this with Spengler's conception of the central role of time in the occidental world picture (cf. p. 234f.), which, from a different viewpoint, comes to the identical conclusion.

However, the—for us—self-evident distinction between past, present, and future does not exist in the Hopi language. It makes no distinction between tenses, but indicates the validity a statement has: fact, memory, expectation, or custom. There is no difference in Hopi between "he runs," "he is running," "he ran," all being rendered by *wari*, "running occur." An expectation is rendered by *warinki* ("running occur [I] daresay"), which covers "he will, shall, should, would run." If, however, it is a statement of a general law, *warikngwe* ("running occur, characteristically") is applied (La Barre, 1954, pp. 197 ff.). The Hopi "has no general notion or intuition of time as a smooth flowing continuum in which everything in the universe proceeds at an equal rate, out of a future, through a present, into a past" (Whorf, 1952, p. 67). Instead of our categories of space and time, Hopi rather distinguishes the "manifest," all that which is accessible to the senses, with no distinction between present and past, and the "unmanifest" comprising the future as well as what we call mental. Navaho (cf. Kluckhohn and Leighton, 1951) has little development of tenses; the emphasis is upon types of activity, and thus it distinguishes durative, perfective, usitative, repetitive, iterative, optative, semifactive, momentaneous, progressive, transitional, conative, etc., aspects of action. The difference can be defined that the first concern of English (and Indo-European language in general) is time, of Hopi—validity, and of Navaho—type of activity (personal communication of Professor Kluckhohn).

Whorf asks:

> How would a physics constructed along these lines work, with no t (time) in its equations? Perfectly, as far as I can see, though of course it would require different ideology and perhaps different mathematics. Of course, v (velocity) would have to go, too (1952, p. 7).

Again, it can be mentioned that a timeless physics actually exists, in the form of Greek statics (cf. p. 234). For us, it is part of a wider system, dynamics, for the particular case that $t \rightarrow \infty$, i.e., time approaches the infinite and drops out of the equations.

i.e., time approaches the infinite and drops out of the equations.

As regards space, the Indo-European tongues widely express nonspatial relations by spatial metaphors: long, short for duration; heavy, light, high, low for intensity; approach, rise, fall for tendency; Latin expressions like *educo, religio, comprehendo* as metaphorical spatial (probably more correct: corporeal, L.v.B.) references: lead out, tying back, grasp, etc.[2]

This is untrue of Hopi where rather physical things are named by psychological metaphors. Thus the Hopi word for "heart" can be shown to be a late formation from a root meaning "think" or "remember." Hopi language is, as Whorf states, capable of accounting for and describing correctly, in a pragmatical or observational sense, all observable phenomena of the universe. However, the implicit metaphysics is entirely different, being rather a way of animistic or vitalistic thinking, near to the mystical experience of oneness.

Thus, Whorf maintains, "Newtonian space, time and matter are no intuitions. They are recepts from culture and language." (1952, p. 40).

> Just as it is possible to have any number of geometries other than the Euclidean which give an equally perfect account of space configurations, so it is possible to have descriptions of the universe, all equally valid, that do not contain our familiar contrast of time and space. The relativity viewpoint of modern physics is one such view, conceived in mathematical terms, and the Hopi Weltanschauung is another and quite different one, non-mathematical and linguistic (Whorf, 1952, p. 67).

The ingrained mechanistic way of thinking which comes into difficulties with modern scientific developments is a consequence of our specific linguistic categories and habits, and Whorf hopes that insight into the diversity of linguistic systems may contribute to the reevaluation of scientific concepts.

La Barre (1954, p. 301) has vividly summarized this viewpoint:

> Aristotelian Substance and Attribute look remarkably like Indo-European nouns and predicate adjectives. . . . More modern science may well raise the question whether Kant's Forms, or twin "spectacles" of Time and Space (without which we can perceive nothing) are not on the one hand mere Indo-European verbal tense, and on the other hand human stereoscopy and kinaesthesis and life-process—which might be more economically expressed in terms of the c, or light-constant, of Einstein's formula. But we must remember all the time that $E = mc^2$ is also only a grammatical conception of reality in terms of Indo-European morphological categories of speech. A Hopi, Chinese, or Eskimo Einstein might discover via his grammatical habits wholly different mathematical conceptualizations with which to apperceive reality.

This paper is not intended to discuss the linguistic problems posed by Whorf as was exhaustively done in a recent symposium

(Hoijer *et al.*, 1954). However, it has occured to the present author that what is known as the Whorfian hypothesis is not an isolated statement of a somewhat extravagant individual. Rather the Whorfian hypothesis of the linguistic determination of the categories of cognition is part of a general revision of the cognitive process. It is embedded in a powerful current of modern thought, the sources of which can be found in philosophy as well as in biology. It seems that these connections are not realized to the extent they deserve.

The general problem posed may be expressed as follows: In how far are the categories of our thinking modeled by and dependent on biological and cultural factors? It is obvious that, stated in this way, the problem far exceeds the borders of linguistics and touches the question of the foundations of human knowledge.

Such analysis will have to start with the classical, absolutistic world view which found its foremost expression in the Kantian system. According to the Kantian thesis, there are the so-called forms of intuition, space and time, and the categories of the intellect, such as substance, causality and others which are universally committal for any rational being. Accordingly science, based upon these categories, is equally universal. Physical science using these a priori categories, namely, Euclidean space, Newtonian time and strict deterministic causality, is essentially classical mechanics which, therefore, is the absolute system of knowledge, applying to any phenomenon as well as to any mind as observer.

It is a well-known fact that modern science has long recognized that this is not so. There is no need to belabor this point. Euclidean space is but one form of geometry beside which other, non-Euclidean geometries exist which have exactly the same logical structure and right to exist. Modern science applies whatever sort of space and time is most convenient and appropriate for describing the events in nature. In the world of medium dimensions, Euclidean space and Newtonian time apply in the way of satisfactory approximations. However, coming to astronomical dimensions and, on the other hand, to atomic events, non-Euclidean spaces or the many-dimensional configuration spaces of quantum theory are introduced. In the theory of relativity, space and time fuse in the Minkowski union, where time is another coordinate of a four-dimensional continuum, although of a somewhat peculiar character. Solid matter, this most obtrusive part of experience and most trivial of the categories of naïve physics, consists almost completely of holes, being a void for the greatest part, only interwoven by centers of energy which, considering their magnitude, are separated by astronomical distances. Mass and energy, somewhat sophisticated quantifications of the categorical antithesis of stuff and force, appear as expressions of one unknown reality, interchangeable according to Einstein's law. Similarly, the strict determinism of classical physics is replaced in quantum physics by indeterminism or rather by the insight that the laws of nature are essentially of a statistical character. Little is left of Kant's supposedly a priori and absolute categories. Incidentally, it is symptomatic of the relativity of world views that Kant who, in his epoch, appeared to be the great destroyer of all "dogmatism," to us appears a paradigm of unwarranted absolutism and dogmatism.

So the question arises—what is it which determines the categories of human cognition? While, in the Kantian system, the categories appeared to be absolute for any rational observer, they now appear as changing with the advancement of scientific knowledge. In this sense, the absolutistic conception of earlier times and of classical physics is replaced by a scientific relativism.

The argument of the present discussion may be defined as follows. The categories of knowledge, of everyday knowledge as well as of scientific knowledge, which in the last resort is only a refinement of the former, depend, first, on biological factors; second, on cultural factors. Third, notwithstanding this all-too-human entanglement, absolute knowledge, emancipated from human limitations, is possible in a certain sense.

The Biological Relativity of Categories

Cognition is dependent, firstly, on the psycho-physical organization of man. We may refer here in particular to that approach in modern biology which was inaugurated by Jacob von Uexküll under the name of *Umwelt-Lehre*. It essentially amounts to the statement that, from the great cake of reality, every living organism cuts a slice, which it can perceive and to which it can react owing to its psycho-physical organization, i.e., the structure of receptor and effector organs. Von Uexküll and Kriszat (1934) have presented fascinating pictures how the same section of nature looks as seen by various animals; they should be compared to Whorf's equally amusing drawings which show how the world is modeled according to linguistic schemes. Here only a few examples, chosen from Uexküll's extensive behavioral studies, can be mentioned.

Take, e.g., a unicellular organism like the paramecium. Its almost only way of response is the flight reaction (phobotaxis) by which it reacts to the most diverse, chemical, tactile, thermal, photic, etc., stimuli. This simple reaction, however, suffices safely to guide that animal which possesses no specific sense organs, into the region of optimal conditions. The many things in the environment of the paramecium, algae, other infusoria, little crustaceans, mechanical obstacles and the like, are nonexistent for it. Only one stimulus is received which leads to the flight reaction.

As this example shows, the organizational and functional plan of a living being determines what can become "stimulus" and "characteristic" to which the organism responds with a certain reaction. According to von Uexküll's expression, any organism, so to speak, cuts out from the multiplicity of surrounding objects a small number of characteristics to which it reacts and whose ensemble forms its "ambient" (*Umwelt*). All the rest is nonexistent for that particular organism. Every animal is surrounded, as by a soapbubble, by its specific ambient, replenished by those characteristics which are amenable to it. If, reconstructing an animal's ambient, we enter this soapbubble, the world is profoundly changed: Many characteristics disappear, others arise, and a completely new world is found.

Von Uexküll has given innumerable examples delineating the ambients of various animals. Take, for instance, a tick lurking in the bushes for a passing mammal in whose skin it settles and drinks itself full of blood. The signal is the odor of butyric acid, flowing from the dermal glands of all mammals. Following this stimulus, it plunges down; if it fell on a warm body—as monitored by its sensitive thermal sense—it has reached its prey, a warm-blooded animal, and only needs to find, aided by tactile sense, a hair-free place to pierce in. Thus the rich environment of the tick shrinks to metamorphize into a scanty configuration out of which only three signals, beaconlike, are gleaming which, however, suffice to lead the animal surely to its goal. Or again, some sea urchins respond to any darkening by striking together their spines. This reaction invariably is applied against a passing cloud or boat, or the real enemy, an approaching fish. Thus, while the environment of the sea urchin contains many different objects, its ambient only contains one characteristic, namely, dimming of light.

This organizational constraint of the ambient goes even much farther than the examples just mentioned indicate (von Bertalanffy, 1937). It also concerns the forms of intuition, considered by Kant as a priori and immutable. The biologist finds that there is no absolute space or time but that they depend on the organization of the perceiving organism. Three-dimensional Euclidean space, where the three rectangular coordinates are equivalent, was always identified with the a priori space of experience and perception. But even simple contemplation shows, and experiments in this line (von Allesch, 1931; von Skramlik, 1934, and others) prove that the space of visual and tactual perception is in no way Euclidean. In the space of perception, the coordinates are in no way equivalent, but there is a fundamental difference between top and bottom, right and left, and fore and aft. Already the organization of our body and, in the last resort, the fact that the organism is subjected to gravity, makes for an inequality of the horizontal and vertical dimensions. This is readily shown by a simple fact known to every photographer. We experience it as quite correct that, according to the laws of perspective, parallels, such as railroad tracks, converge in the distance. Exactly the same perspective foreshortening is, however, experienced as being wrong if it appears in the vertical dimension. If a picture was taken with the camera tilted, we obtain "falling lines," the edges of a house, e.g., running together. This is, perspectively, just as correct as are the converging railroad tracks; nevertheless, the latter perspective is experienced as being correct, while the converging edges of a house are experienced as wrong; the explanation being that the human organism is such as to have an ambient with considerable horizontal, but negligible vertical extension.[3]

A similar relativity is found in experienced time. Von Uexküll has introduced the notion of the "instant" as the smallest unit of perceived time. For man, the instant is about 1/18 sec., i.e., impressions shorter than this duration are not perceived separately but fuse. It appears that the duration of the instant depends not on conditions in the sense organs but rather in the central nervous system, for it is the same for different sense organs. This flicker fusion is, of course, the *raison d'être* of movie pictures when frames presented in a sequence faster than 18 per second fuse into continuous motion. The duration of the instant varies in different species. There are "slow motion-picture animals" (von Uexküll) which perceive a greater number of impressions per second than man. Thus, the fighting fish (*Betta*) does not recognize its image in a mirror if, by a mechanical device, it is presented 18 times per second. It has to be presented at least 30 times per second; then the fish attacks his imaginary opponent. Hence, these small and very active animals consume a larger number of impressions than man does, per unit of astronomical time; time is decelerated. Conversely, a snail is a "rapid motion-picture animal." It crawls on a vibrating stick if it approaches four times per second, i.e., a stick vibrating four times per second appears at rest to the snail.

Experienced time is not Newtonian. Far from flowing uniformly (*aequilabiliter fluit*, as Newton has it), it depends on physiological conditions. The so-called time memory of animals and man seems to be determined by a "physiological clock." Thus bees, conditioned to appear at a certain time at the feeding place, will show up earlier or later if drugs which increase or decrease the rate of metabolism are administered (e.g., von Stein-Beling, 1935; Kalmus, 1934; Wahl, 1932; and others).

Experienced time seems to fly if it is filled with impressions, and creeps if we are in a state of tedium. In fever, when body temperature and metabolic rate are increased, time seems to linger since the number of "instants" per astronomical unit in Uexküll's sense is increased. This time experience is paralleled by a corresponding increase of the frequency of the α-waves in the brain (Hoagland, 1951). With increasing age, time appears to run faster, i.e., a smaller number of instants is experienced per astronomical unit of time. Correspondingly, the rate of cicatrization of wounds is decreased proportional to age, the psychological as well as physiological phenomena obviously being connected with the slowing-down of metabolic processes in senescence (du Noüy, 1937).

Several attempts (Brody, 1937; Backman, 1940; von Bertalanffy, 1951, p. 346) have been made to establish a biological as compared to astronomical time. One means is the homologization of growth curves: If the course of growth in different animals is expressed by the same formula and curve, the units of the time scale (plotted in astronomical time) will be different, and important physiological changes presumably will appear at corresponding points of the curve. From the standpoint of physics, a thermodynamic time, based upon the second principle and irreversible processes, can be introduced as opposed to astronomical time (Prigogine, 1947). Thermodynamic time is nonlinear but logarithmic since it depends on probabilities; it is, for the same reason, statistical; and it is local because determined by the events at a certain point. Probably biological time bears an intimate although by no means simple relation to thermodynamic time.

How the categories of experience depend on physiological states, is also shown by the action of drugs. Under the influence of mescaline, e.g., visual impressions are intensified, and the perception of space and time undergoes profound changes (cf. Anschütz, 1953; A. Huxley, 1954). It would make a most interesting study to investigate the categories of schizophrenics, and it would probably be found that they differ considerably from those of "normal" experience, as do indeed the categories in the experience of dreams.

Even the most fundamental category of experience, namely, the distinction of ego and nonego, is not absolutely fixed. It seems gradually to evolve in the development of the child. It is essentially different in the animistic thinking of the primitives (still in force even in the Aristotelian theory where everything "seeks" its natural place), and in Western thinking since the Renaissance which "has discovered the inanimate" (Schaxel, 1923). The object-subject separation again disappears in the empathic world view of the poet, in mystical ecstasy and in states of intoxication.

There is no intrinsic justification to consider as "true" representation of the world what we take to be "normal" experience (i.e., the experience of the average adult European of the twentieth century), and to consider all other sorts of experience that are equally vivid, as merely abnormal, fantastic or, at best, a primitive precursor to our "scientific" world picture.

The discussion of these problems could easily be enlarged, but the point important for the present topic will have become clear. The categories of experience or forms of intuition, to use Kant's term, are not a universal a priori, but rather they depend on the psychophysical organization and physiological conditions of the experiencing animal, man included. This relativism from the biological standpoint is an interesting parallel to the relativism of categories as viewed from the standpoint of culture and language.

The Cultural Relativity of Categories

We now come to the second point, the dependence of categories on cultural factors. As already mentioned, the Whorfian thesis of the dependence of categories on linguistic factors is part of a general conception of cultural relativism which has developed in the past 50 years; although even this is not quite correct, since Wilhelm von Humboldt has already emphasized the dependence of our world perspective on linguistic factors and the structure of language.

It appears that this development started in the history of art. At the beginning of this century, the Viennese art-historian, Riegl, published a very learned and tedious treatise on late-Roman artcraft. He introduced the concept of *Kunstwollen*, a term which may be translated as "artistic intention." The unnaturalistic character of primitive art was conceived to be a consequence not of a lack of skill or know-how, but rather as expression of an artistic intention which is different from ours, being not interested in a realistic reproduction of nature. The same applies to the so-called degeneration of classic art in the late Hellenistic period. This conception later was expanded by Worringer who demonstrated in the example of Gothic art that artistic modes diametrically opposed to the classical canon are an outcome not of technical impotence, but rather of a different world view. It was not that Gothic sculptors and painters did not know how to represent nature correctly, but their intention was different and not directed towards representative art. The connection of these theories with the primitivism and expressionism in modern art needs no discussion.

I wish to offer another example of the same phenomenon which is instructive since it has nothing to do with the antithesis of representative and expressionistic, objective and abstract art. It is found in the history of the Japanese woodcut.

Japanese pictures of the later period apply a certain kind of perspective, known as parallel perspective, which is different from central perspective as used in European art since the Renaissance. It is well known that Dutch treatises on perspectives were introduced into Japan in the late eighteenth century, and were eagerly studied by the *Ukiyoye* (woodcut) masters. They adopted perspective as a powerful means to represent nature, but only to a rather subtle limit. While European painting uses central perspective where the picture is conceived from a focal point and consequently parallels converge in the distance, the Japanese only accepted parallel perspective; i.e., a way of projection where the focal point is in the infinite, and hence parallels do not converge. We may be sure that this was not lack of skill in those eminent Japanese artists who, like Hokusai and Hiroshige, later exerted a profound influence on modern European art. They certainly would have found no difficulty to adopt an artistic means which even was handed to them ready-made. Rather we may conjecture that they felt central perspective, dependent on the standpoint of the observer, to be contingent and accidental and not representing reality since it changes as the observer moves from one place to another. In a similar way, the Japanese artists never painted shadows. This, of course, does not mean that they did not see shadow or go into the shade when the sun was burning. However, they did not wish to paint it; for the shadow does not belong to the reality of things but is only changing appearance.

So the categories of artistic creation seem to be dependent on the culture in question. It is well known that Spengler has expanded this thesis to include cognitive categories. According to him, the so-called a priori contains, besides a small number of universally human and logically necessary forms of thinking, also forms of thinking that are universal and necessary not for humanity as a whole but only for the particular civilization in

question. So there are various and different "styles of cognition,"
characteristic of certain groups of human beings. Spengler does
not deny the universal validity of the formal laws of logic or
of the empirical *verités de fait*. He contends, however, the relativity of the contentual a prioris in science and philosophy. It is
in this sense that Spengler states the relativity of mathematics and
mathematical science. The mathematical formulae as such carry
logical necessity; but their visualizable interpretation which gives
them meaning is an expression of the "soul" of the civilization
which has created them. In this way, our scientific world picture
is only of relative validity. Its fundamental concepts, such as the
infinite space, force, energy, motion, etc., are an expression of
our occidental type of mind, and do not hold for the world
picture of other civilizations.

The analysis upon which Spengler's cultural relativism of the
categories is mainly based is his famous antithesis of Apollonian
and Faustian man. According to him, the primeval symbol of the
Apollonian mind of antiquity is the material and bodily existence
of individuals; that of the Faustian mind of the occident is infinite
space. Thus "space," for the Greeks, is the *mè ón*, that which is
not. Consequently, Apollonian mathematics is a theory of visualizable magnitudes, culminating in stereometry and geometric
construction which, in occidental mathematics, is a rather inconsequential elementary topic. Occidental mathematics, governed
by the primeval symbol of the infinite space, in contrast, is a
theory of pure relations, culminating in differential calculus, the
geometry of many-dimensional spaces, etc., which, in their un-
visualizability, would have been completely inconceivable to
the Greeks.

A second antithesis is that of the static character of Greek,
and the dynamic character of occidental thought. Thus, for the
Greek physicist, an atom was a miniature plastic body; for
occidental physics, it is a center of energy, radiating actions into
an infinite space. Connected with this is the meaning of time.
Greek physics did not contain a time dimension, and this is at
the basis of its being a statics. Occidental physics is deeply concerned with the time course of events, the notion of entropy
being probably the deepest conception in the system. From the
concern with time further follows the historical orientation of the
occidental mind expressed in the dominating influence of the

clock, in the biography of the individual, in the enormous perspective of "world history" from historiography to cultural history to anthropology, biological evolution, geological history, and finally astronomical history of the universe. Again, the same contrast is manifest in the conception of the mind. Static Greek psychology imagines a harmonic soul-body whose "parts," according to Plato, are reason (*logistikón*), emotion (*thymoeidés*), and cathexis (*epithymetikón*). Dynamic occidental psychology imagines a soul-space where pyschological forces are interacting.

Taking exception from Spengler's metaphysics and intuitive method, and disregarding questionable details, it will be difficult to deny that his conception of the cultural relativity of categories is essentially correct. It suffices to remember the first lines of the Iliad, telling of the heroes of the Trojan war *autoùs te helória teûche kýnessin*, that their selves were given a prey to the hounds and birds, the "self" being essentially the body or *sŏma*. Compare this with Descartes' *cogito ergo sum*—and the contrast between Apollonian and Faustian mind is obvious.

While the German philosophers of history were concerned with the small number of high cultures (*Hochkulturen*), it is the hallmark and merit of modern and, in particular, American anthropology to take into account the entire field of human "cultures" including the multiplicity exhibited by primitive peoples. So the theory of cultural relativism wins a broader basis but it is remarkable that the conclusions reached are very similar to those of the German philosophers. In particular, the Whorfian thesis is essentially identical with the Spenglerian—the one based upon the linguistics of primitive tribes, the other on a general view of the few high cultures of history.[4]

So it appears well established that the categories of cognition depend, first, on biological factors, and secondly, on cultural factors. A suitable formulation perhaps can be given in the following way.

Our perception is essentially determined by our specifically human, psychophysical organization. This is essentially von Uexküll's thesis. Linguistic, and cultural categories in general, will not change the potentialities of sensory experience. They will, however, change apperception, i.e., which features of experienced reality are focused and emphasized, and which are underplayed.

There is nothing mysterious or particularly paradoxical in this statement which, on the contrary, is rather trivial; nothing which would justify the heat and passion which has often characterized the dispute on the Whorfian, Spenglerian, and similar theses. Suppose a histological preparation is studied under the microscope. Any observer, if he is not color-blind, will perceive the same picture, various shapes and colors, etc., as given by the application of histological stains. However, what he actually sees, i.e., what is his apperception (and what he is able to communicate), depends widely on whether he is an untrained or a trained observer. Where for the layman there is only a chaos of shapes and colors, the histologist sees cells with their various components, different tissues, and signs of malignant growth. And even this depends on his line of interest and training. A cytochemist will possibly notice fine granulations in the cytoplasm of cells which represent to him certain chemically defined inclusions; the pathologist may, instead, entirely ignore these niceties, and rather "see" how a tumor has infiltrated the organ. Thus what is seen depends on our apperception, on our line of attention and interest which, in turn, is determined by training, i.e., by linguistic symbols by which we represent and summarize reality.

It is equally trivial that the same object is something quite different if envisaged from different viewpoints. The same table is to the physicist an aggregate of electrons, protons, and neutrons, to the chemist a composition of certain organic compounds, to the biologist a complex of wood cells, to the art historian a baroque object, to the economist a utility of certain money value, etc. All these perspectives are of equal status, and none can claim more absolute value than the other (cf. von Bertalanffy, 1953b). Or, take a slightly less trivial example. Organic forms can be considered from different viewpoints. Typology considers them as the expression of different plans of organization; the theory of evolution as a product of a historical process; a dynamic morphology as expression of a play of processes and forces for which mathematical laws are sought (von Bertalanffy, 1941). Each of these viewpoints is perfectly legitimate, and there is little point to play one against the other.

What is obvious in these special examples equally holds for what traits of reality are noticed in our general world picture. It is an important trend of the development of science that new aspects, previously unnoticed, are "seen," i.e., come under the focus of attention and apperception; and conversely, an important obstacle that the goggles of a certain theoretical conception do not allow to realize phenomena which, in themselves, are perfectly obvious. History of science is rich in examples of such kind. For instance, the theoretical spectacles of a one-sided "cellular pathology" simply did not allow one to see that there are regulative relations in the organism as a whole which is more than a sum or aggregate of cells; relations which were known to Hippocrates and have found a happy resurrection in the modern doctrine of hormones, of somatotypes and the like. The modern evolutionist, guided by the theory of random mutation and selection, does not see that an organism is obviously more than a heap of hereditary characteristics or genes shuffled together by accident. The mechanistic physicist did not see the so-called secondary qualities like color, sound, taste, etc., because they do not fit into his scheme of abstractions; although they are just as "real" as are the supposedly basic "primary qualities" of mass, impenetrability, motion and the like, the metaphysical status of which is equally dubious, according to the testimony of modern physics.

Another possible formulation of the same situation, but emphasizing another aspect, is this. Perception is universally human, determined by man's psychophysical equipment. Conceptualization is culture-bound because it depends on the symbolic systems we apply. These symbolic systems are largely determined by linguistic factors, the structure of the language applied. Technical language, including the symbolism of mathematics, is, in the last resort, an efflorescence of everyday language, and so will not be independent of the structure of the latter. This, of course, does not mean that the content of mathematics is "true" only within a certain culture. It is a tautological system of hypothetico-deductive nature, and hence any rational being accepting the premises must agree to all its deductions. But which aspects or perspectives are mathematized depends on the cultural context. It is perfectly possible that different individuals and cultures have different predilections for choosing certain aspects and neglecting others.[5] Hence, for example, the Greek's concern with geometrical problems and the concern of occidental mathematics with calculus, as emphasized by Spengler; hence the appearance of unorthodox fields of mathematics, such as topology, group theory, game theory and the like, which do not fit into the popular notion of mathematics as a "science of quantities"; hence the individual physicist's predilection for, say, "macroscopic" classical thermodynamics or "microscopic" molecular statistics, for matrix mechanics or wave mechanics to approach the same phenomena. Or, speaking more generally, the analytic type of mind concerned with what is called "molecular" interpretations, i.e., the resolution and reduction of phenomena to elementaristic components; and the holistic type of mind concerned with "molar" interpretations, i.e., interested in the laws that govern the phenomenon as a whole. Much harm has been done in science by playing one aspect against the other and so, in the elementaristic approach, to neglect and deny obvious and most important characteristics; or, in the holistic approach, to deny the fundamental importance and necessity of analysis.

It may be mentioned, in passing, that the relation between language and world view is not unidirectional but reciprocal, a fact which perhaps was not made sufficiently clear by Whorf. The structure of language seems to determine which traits of reality are abstracted and hence what form the categories of thinking take on. On the other hand, the world outlook determines and forms the language.

A good example is the evolution from classical to medieval Latin. The Gothic world view has recreated an ancient language, this being true for the lexical as well as the grammatical aspect. Thus the scholastics invented hosts of words which are atrocities from the standpoint of Cicero's language (as the humanists of the Renaissance so deeply felt in their revivalistic struggle); words introduced to cope with abstract aspects foreign to the corporeally-thinking Roman mind, like *leonitas, quidditas* and the rest of them. Equally, although the superficial rules of grammar were observed, the line of thinking and construction was profoundly altered. This also applies to the rhetorical aspect, as in the introduction of the end-rhyme in contrast to the classical meters. Comparison, say, of the colossal lines of the *Dies irae* with some Virgilian or Horatian stanza makes obvious not only the tremendous gap between different "world-feelings" but the determination of language by the latter as well.

The Perspectivistic View

Having indicated the biological and cultural relativity of the categories of experience and cognition, we can, on the other hand, also indicate the limits of this relativity, and thus come to the third topic stated in the beginning.

Relativism has often been formulated to express the purely conventional and utilitarian character of knowledge, and with the emotional background of its ultimate futility. We can, however, easily see that such consequence is not implied.

A suitable starting point for such discussion are the views on human knowledge expressed by von Uexküll in connection with his *Umweltlehre* which we have discussed earlier. According to him, the world of human experience and knowledge is one of the innumerable ambients of the organisms, in no way singular as compared to that of the sea urchin, the fly or the dog. Even the world of physics, from the electrons and atoms up to galaxies, is a merely human product, dependent upon the psychophysical organization of the human species.

Such conception, however, appears to be incorrect. This may be shown in view of the levels both of experience and of abstract thinking, of everyday life and of science.

As far as direct experience is concerned, the categories of perception as determined by the biophysiological organization of the species concerned cannot be completely "wrong," fortuitous and arbitrary. Rather they must, in a certain way and to a certain extent, correspond to "reality"—whatever this means in a metaphysical sense. Any organism, man included, is not a mere spectator, looking at the world scene and hence free to adopt spectacles, however distorting, such as the whims of God, of biological evolution, of the "soul" of culture, or of language have put on his metaphorical nose. Rather he is a reactor and actor in the drama. The organism has to react to stimuli coming from outside, according to its innate psychophysical equipment. There is a latitude in what is picked up as a stimulus, signal and characteristic in Uexküll's sense. However, its perception must allow the animal to find its way in the world. This would be impossible if the categories of experience, such as space, time, substance, causality, were entirely deceptive. The categories of experience have arisen in biological evolution, and have continually to justify themselves in the struggle for existence. If they would not, in some way, correspond to reality, appropriate reaction would be impossible, and such organism would quickly be eliminated by selection.

Speaking in anthropomorphic terms: A group of schizophrenics who share their illusions may get along with each other pretty well; they are, however, utterly unfit to react and adapt themselves to real outside situations, and this is precisely the reason why they are put into the asylum. Or, in terms of Plato's simile: the prisoners in the cave do not see the real things but only their shadows; but if they are not only looking at the spectacle, but have to take part in the performance, the shadows must, in some way, be representative of the real things. It seems to be the most serious shortcoming of classic occidental philosophy, from Plato to Descartes and Kant, to consider man primarily as a spectator, as *ens cogitans*, while, for biological reasons, he has essentially to be a performer, an *ens agens* in the world he is thrown in.

Lorenz (1943) has convincingly shown that the "a priori" forms of experience are of essentially the same nature as the innate schemata of instinctive behavior, following which animals respond to companions, sexual partners, offspring or parents, prey or predators, and other outside situations. They are based upon psychophysiological mechanisms, such as the perception of space is based on binocular vision, parallax, the contraction of the ciliary muscle, apparent increase or decrease in size of an approaching or receding object, etc. The "a priori" forms of intuition and categories are organic functions, based upon corporeal and even machine-like structures of the sense organs and the nervous system, which have evolved as adaptations in the millions of years of evolution. Hence they are fitted to the "real" world in exactly the same way and for the same reason, as the equine hoof is fitted to the steppe terrain, the fin of the fish to the water. It is a preposterous anthropomorphism to assume that the human forms of experience are the only possible ones, valid for any rational being. On the other hand, the conception of the forms of experience as an adaptive apparatus, proved in millions of years of struggle for existence, guarantees that there is a sufficient correspondence between "appearance" and "reality." Any stimulus is experienced not as it is but as the organism reacts to it, and thus the world-picture is determined by psychophysical organization. However, where a paramecium reacts with its phobotactic reaction, the human observer, though his world outlook is quite different, also actually finds an obstacle when he uses his microscope. Similarly, it is well possible to indicate which traces of experience correspond to reality, and which, comparable to the colored fringes in the field of a microscope which is not achromatically corrected, do not. So Pilate's question, "What is Truth," is to be answered thus: Already the fact that animals and human beings are still in existence, proves that their forms of experience correspond, to some degree, with reality.

In view of this, it is possible to define what is meant by the intentionally loose expression used above, that experience must correspond "in a certain way" to "reality whatever this means." It is not required that the categories of experience fully correspond to the real universe, and even less that they represent it completely. It suffices—and that is Uexküll's thesis—that a rather small selection of stimuli is used as guiding signals. As for the connections of these stimuli, i.e., the categories of experience, they need not mirror the nexus of real events but must, with a certain tolerance allowed, be isomorphic to it. For the biological reasons mentioned above, experience cannot be completely "wrong" and arbitrary; but, on the other hand, it is sufficient that a certain degree of isomorphism exists between the experienced world and the "real" world, so that the experience can guide the organism in such way as to preserve its existence.

Again, to use a simile: The "red" sign is not identical with the various hazards it indicates, oncoming cars, trains, crossing pedestrians, etc. It suffices, however, to indicate them, and thus "red" is isomorphic to "stop," "green" isomorphic to "go."

Similarly, perception and experience categories need not mirror the "real" world; they must, however, be isomorphic to it to such degree as to allow orientation and thus survival.

But these deductive requirements are precisely what we actually find. The popular forms of intuition and categories, such as space, time, matter and causality, work well enough in the world of "medium dimensions" to which the human animal is biologically adapted. Here, Newtonian mechanics and classical physics, as based upon these visualizable categories, are perfectly satisfactory. They break down, however, if we enter universes to which the human organism is not adapted. This is the case, on the one hand, in atomic dimensions, and in cosmic dimensions on the other.

Coming now to the world of science, Uexküll's conception of the physical universe as but one of the innumerable biological ambients, is incorrect or at least incomplete. Here a most remarkable trend comes in which may be called the progressive de-anthropomorphization of science (von Bertalanffy, 1937, 1953b). It appears that this process of de-anthropomorphization takes place in three major lines.

It is an essential characteristic of science that it progressively de-anthropomorphizes, that is, progressively eliminates those traits which are due to specifically human experience. Physics necessarily starts with the sensory experience of the eye, the ear, the thermal sense, etc., and thus builds up fields like optics, acoustics, theory of heat, which correspond to the realms of sensory experience. Soon, however, these fields fuse into such that do not have any more relation to the "visualizable" or "intuitable": Optics and electricity fuse into electromagnetic theory, mechanics and theory of heat into statistical thermodynamics, etc.

This evolution is connected with the invention of artificial sense-organs and the replacement of the human observer by the recording instrument. Physics, though starting with everyday experience, soon transgresses it by expanding the universe of experience through artificial sense organs. Thus, for example, instead of seeing only visible light with a wave length between 380 and 760 millimicra, the whole range of electromagnetic radiation, from shortest cosmic rays up to radio waves of some kilometers wave length, is disclosed.

Thus it is one function of science to expand the observable. It is to be emphasized that, in contrast to a mechanistic view, we do not enter another metaphysical realm with this expansion. Rather the things surrounding us in everyday experience, the cells seen in a microscope, the large molecules observed by the electron microscope, and the elementary particles "seen," in a still more indirect and intricate way, by their traces in a Wilson chamber, are not of a different degree of reality. It is a mechanistic superstition to believe that atoms and molecules (speaking with Alice in the Wonderland of Physics) are "realer" than apples, stones and tables. The ultimate particles of physics are not a metaphysical reality behind observation; they are an expansion of what we observe with our natural senses, by way of introducing suitable artificial sense organs.

In any way, however, this leads to an elimination of the limitations of experience as imposed by the specifically human psychophysical organization, and, in this sense, to the de-anthropomorphization of the world picture.

A second aspect of this development is what is called the convergence of research (cf. Bavink, 1949). The constants of physics have often been considered as only conventional means for the most economic description of nature. The progress of research, however, shows a different picture. First, natural constants such as the mechanical equivalent of heat or the charge of electrons vary widely in the observation of individual observers. Then, with the refinement of techniques, a "true" value is approached asymptotically so that consecutive determinations alter the established value only in progressively smaller digits of decimals. Not only this: Physical constants such as Loschmidt's number and its like are established not by one method but perhaps by 20 methods which are completely independent of each other. In this way, they cannot be conceived as being simply conventions for describing phenomena economically; they represent certain aspects of reality, independent of biological, theoretical or cultural biases. It is indeed one of the most important occupations of natural science thus to verify its findings in mutually independent ways.

However, perhaps the most impressive aspect of progressive de-anthropomorphization is the third. First, the so-called secondary qualities go, that is, color, sound, smell, taste disappear from the physical world picture since they are determined by the so-called specific energy of the diverse and specifically human senses. So, in the world picture of classical physics, only the primary qualities such as mass, impenetrability, extension, etc., are left which, psychophysically, are characterized as being the common ground of visual, tactual, acoustical experience. Then, however, these forms of intuition and categories also are eliminated as being all-too-human. Even Euclidean space and Newtonian time of classical physics, as was noted previously, are not identical with the space and time of direct experience; they already are constructs of physics. This, of course, is true even more of the theoretical structures of modern physics.

Thus, what is specific of our human experience is progressively eliminated. What eventually remains is only a system of mathematical relations.

Some time ago it was considered a grave objection against the theory of relativity and quantum theory that it became increasingly "unvisualizable," that its constructs cannot be represented by imaginable models. In truth, however, this is a proof that the system of physics detaches itself from the bondage of our specifically human sensory experience; a pledge that the system of physics in its consummate form—leaving it undecided whether this is attained or even is attainable at all—does not belong to the human ambient (*umwelt* in Uexküll's sense) any more but is universally committal.

In a way, progressive de-anthropomorphization is like Muenchhausen pulling himself out of the quagmire on his own pigtail. It is, however, possible because of a unique property of symbolism. A symbolic system, an algorithm, such as that of mathematical physics, wins a life of its own as it were. It becomes a thinking machine, and once the proper instructions are fed in, the machine runs by itself, yielding unexpected results that surpass the initial amount of facts and given rules, and are thus unforeseeable by the limited intellect who originally has created the machine. In this sense, the mechanical chess player can outplay its maker (Ashby, 1952a), i.e., the results of the automatized symbolism transcend the original input of facts and instructions. This is the case in any algorithmic prediction, be it a formal deduction on any level of mathematical difficulty or a physical prediction like that of still unknown chemical elements or planets (cf. von Bertalanffy, 1956a). Progressive de-anthropomorphization, that is, replacement of direct experience by a self-running algorithmic system, is one aspect of this state of affairs.

Thus, the development of physics naturally depends on the psychophysical constitution of its creators. If man would not perceive light but radium or x-rays which are invisible to us, not only the human ambient but also the development of physics would have been different. But in a similar way, as we have discovered, by means of suitable apparatus and supplementing our sensory experience, x-rays and all the range of electromagnetic radiations, the same would be true of beings with an entirely different psychophysical constitution. Suppose there are intelligent beings or "angels" on a planet of the Sirius who perceive only x-rays; they would have detected, in a corresponding way, those wave lengths that mean visible light to us. But not only this: The Sirius angels would possibly calculate in quite different systems of symbols and theories. However, since the system of physics, in its consummate state, does not contain anything human any more, and the corresponding thing would be true of any system of physics, we must conclude that those physics, although different in their symbolic systems, have the same content, that is, the mathematical relations of one physics could be translated by means of a suitable "vocabulary" and "grammar" into those of the other.

This speculation is not quite utopian, but, to a certain extent, seen in the actual development of physics. Thus, classical thermodynamics and molecular statistics are different "languages" using different abstractions and mathematical symbolisms, but the statements of one theory can readily be translated into the other. This even has quite timely implications; thermodynamics and the modern theory of information obviously are similarly isomorphic systems, and the elaboration of a complete "vocabulary" for translation is in progress.

If, in the sense just indicated, the system of physics in its ideal state, which can be approached only asymptotically, is absolute, we must, however, not forget another and in some way antithetical aspect. What traits of reality we grasp in our theoretical system is arbitrary in the epistemological sense, and determined by biological, cultural and probably linguistic factors.

This, again, has first a trivial meaning. The Eskimos are said to have some 30 different names for "snow," doubtless because it is vitally important for them to make fine distinctions while, for us, differences are negligible. Conversely, we call machines which are only superficially different, by the names of Fords, Cadillacs, Pontiacs and so forth, while for the Eskimos they would be pretty much the same. The same, however, is true in a non-trivial sense, applying to general categories of thinking.

It would be perfectly possible that rational beings of another structure choose quite different traits and aspects of reality for building theoretical systems, systems of mathematics and physics. Our main concern, probably determined by the grammar of Indo-European language, is with measurable qualities, isolable units, and the like. Our physics neglects the so-called primary qualities; they come in only rudimentarily in the system of physics or in certain abstractions of physiological optics like the color cycle or triangle.[6] Similarly, our way of thinking is conspicuously unfit for dealing with problems of wholeness and form. Therefore, it is only with the greatest effort that holistic as contrasted to elementalistic traits can be included—although they are not less "real." The way of thinking of occidental physics leaves us on the spot if we are confronted with problems of form —hence this aspect, predominant in things biological, is but a tremendous embarrassment to physics.

It may well be that quite different forms of science, of mathematics in the sense of hypothetico-deductive systems, are possible for beings who don't carry our biological and linguistic constraints; mathematical "physics" that are much more fit than ours to deal with such aspects of reality.

The same seems even to be true of mathematical logic. So far, it seems to cover only a relatively small segment of what can easily be expressed in vernacular or mathematical language. The Aristotelian logic, for millenia considered as giving the general and supreme laws of reasoning, actually covers only the extremely small field of subject-predicate relations. The all-or-none concepts of traditional logic fall short of continuity concepts basic for mathematical analysis (cf. von Neumann, 1951, p. 16). Probably it is only a very small field of possible deductive reasoning which is axiomatized even by the efforts of modern logicians.

It may be that the structure of our logic is essentially determined by the structure of our central nervous system. The latter is essentially a digital computer, since the neurons work according to the all-or-nothing law of physiology in terms of yes-or-no decisions. To this corresponds the Heraclitean principle of our thinking in opposites, our bivalent yes-or-no logic, Boolean algebra, and the binary system of numbers[7] to which also the practically more convenient decadic system can be reduced (and is actually reduced in modern calculating machines). Supposing that a nervous system were constructed not after the digital type but as an analog computer (such as, e.g., a slide rule), it may be imagined that a quite different logic of continuity, in contrast to our yes-or-no logic, would arise.

Thus we come to a view which may be called perspectivism (cf. von Bertalanffy, 1953b). In contrast to the "reductionist" thesis that physical theory is the only one to which all possible science and all aspects of reality eventually should be reduced, we take a more modest view. The system of physics is committal for any rational being in the sense explained; that is, by a process of de-anthropomorphization it approaches a representation of certain relational aspects of reality. It is essentially a symbolic algorithm suitable for the purpose. However, the choice of the symbolisms we apply and consequently the aspects of reality we represent, depend on biological and cultural factors. There is nothing singular or particularly sacred about the system of physics. Within our own science, other symbolic systems, such as those of taxonomy, of genetics or the history of art, are equally legitimate although they are far from having the same degree of precision. And in other cultures of human beings and among non-human intelligences, basically different kinds of "science" may be possible which would represent other aspects of reality as well or even better than does our so-called scientific world picture.

There is, perhaps, a deep-lying reason why our mental representation of the universe always mirrors only certain aspects or perspectives of reality. Our thinking, at least in occidental but possibly in any human language, is essentially in terms of opposites. As Heraclitus has it, we are thinking in terms of warm and cold, black and white, day and night, life and death, being and becoming. These are naive formulations. But it appears that also the constructs of physics are such opposites, and that for this very reason prove inadequate in view of reality, certain relations of which are expressed in the formulas of theoretical physics. The popular antithesis between motion and rest becomes meaningless in the theory of relativity. The antithesis of mass and energy is superseded by Einstein's conservation law which accounts for their mutual transformation. Corpuscle and wave are both legitimate and complementary aspects of physical reality which, in certain phenomena and respects, is to be described in one way, in others in the second. The contrast between structure and process breaks down in the atom as well as in the living organism whose structure is at the same time the expression and the bearer of a continuous flow of matter and energy. Perhaps the age-old problem of body and mind is of a similar nature, these being different aspects, wrongly hypostatized, of one and the same reality.

All our knowledge, even if de-anthropomorphized, only mirrors certain aspects of reality. If what has been said is true, reality is what Nicholas of Cusa (cf. von Bertalanffy, 1928b) called the *coincidentia oppositorum*. Discoursive thinking always represents only one aspect of ultimate reality, called God in Cusa's terminology; it can never exhaust its infinite manifoldness. Hence ultimate reality is a unity of opposites; any statement holds from a certain viewpoint only, has only relative validity, and must be supplemented by antithetic statements from opposite points of view.

Thus, the categories of our experience and thinking appear to be determined by biological as well as cultural factors. Secondly, this human bondage is stripped by a process of progressive de-anthropomorphization of our world picture. Thirdly, even though de-anthropomorphized, knowledge only mirrors certain aspects or facets of reality. However, fourthly, *ex omnibus partibus relucet totum*, again to use Cusa's expression: Each such aspect has, though only relative, truth. This, it seems, indicates the limitation as well as the dignity of human knowledge.

Notes

1. This and other examples in Whorf's argument are criticized by Whatmough (1955). "As Brugmann showed (*Syntax des einfachen Satzes*, 1925, pp. 17–24), *fulget, pluit, tonat* are simple old ti-stems (nouns 'lightning there, rain there, thunder there') and Whorf was quite wrong when he said that *tonat* (he used that very word) is structurally and logically unparalleled in Hopi." Similarly, "the Hopi for 'prepare,' we are told, is 'to try-for, to practise-upon.' But this is exactly prae-paro." "It will not do to say that Hopi physics could not have had concepts such as space, velocity, and mass, or that they would have been very different from ours. Hopi has no physics because the Hopi are hindered by taboo or magic from experimental investigation." Although one has to surrender to the linguist's authority, it seems amply demonstrated that the style of thinking is different in the several civilizations even though Whorf's supposition that this is more or less solely due to linguistic factors, is open to criticism.

2. It is interesting to note that exactly the same viewpoint was stated by Lorenz (1943) in terms of the biological determination of categories: "The terms which language has formed for the highest functions of our rational thinking still bear so clearly the stamp of their origin that they might be taken from the 'professional language' of a chimpanzee. We 'win insight' into intricate connections, just as the ape into a maze of branches, we found no better expression for our most abstract ways to achieve goals than 'method,' meaning detour. Our tactile space still has, as it were from time to non-jumping lemurs, a particular preponderance over the visual. Hence we have 'grasped' (*erfasst*) a 'connection' (*Zusammenhang*) only if we can 'comprehend' (*begreifen*, i.e., seize) it. Also the notion of object (*Gegenstand*, that which stands against us) originated in the haptic perception of space. . . . Even time is represented, for good or wrong, in terms of the visualizable model of space (p. 344). . . . Time is absolutely unvisualizable and is, in our categorical thinking, made visualizable always [?; perhaps a Western prejudice, L. v. B.] only by way of spatio-temporal processes. . . . The 'course of time' is symbolized, linguistically and certainly also conceptually, by motion in space (the stream of time). Even our prepositions 'before' and 'after,' our nouns 'past, present and future' originally have connotations representing spatio-temporal configurations of motion. It is hardly possible to eliminate from them the element of motion in space" (pp. 351 ff.).

3. As far as can be seen, this simple demonstration of the non-Euclidean structure of the visual space was first indicated by von Bertalanffy (1937, p. 155), while "curious enough, no reference whatever is found in the literature on the physiology of perception" (Lorenz, 1943, p. 335).

4. An excellent analysis on the culture-dependence of perception, cognition, affect, evaluation, unconscious processes, normal and abnormal behavior, etc., is given in Kluckhohn (1954). The reader is referred to this paper for ample anthropological evidence.

5. I find that Toynbee (1954, pp. 699 ff.), in his otherwise not overly friendly comment on Spengler's theory of types of mathematical thinking, arrives at an identical formulation. He speaks of a different "penchant" of civilizations for certain types of mathematical reasoning, which is the same as the above-used notion of "predilection." The present writer's interpretation of Spengler was, in the essentials, given in 1924, and he has seen no reason to change it.

6. This perhaps can lead to a fairer interpretation of Goethe's "Theory of Colors." Goethe's revolt against Newtonian optics which is a scandal and completely devious within the history of occidental physics, can be understood in this way: Goethe, an eminently eidetic and intuitive mind, had the feeling (which is quite correct) that Newtonian optics purposely neglects, and abstracts from, exactly those qualities which are most prominent in sensory experience. His

Farbenlehre, then, is an attempt to deal with those aspects of reality which are not covered by conventional physics; a theoretical enterprise which remained abortive.

7. Notice the theological motive in Leibniz's invention of the binary system. It represented Creation since any number can be produced by a combination of "something" (1) and "nothing" (0). But has this antithesis metaphysical reality, or is it but an expression of linguistic habits and of the mode of action of our nervous system?

Appendix I: Notes on Developments in Mathematical System Theory (1971)

The program of mathematical system theory has materialized in recent years; mathematical system theory has become an extensive and rapidly growing field. This development is due, on the one hand, to the theoretical problems of "system" as such and in relation to other disciplines; on the other, to problems of the technology of control and communication.

No systematic treatment or comprehensive review of mathematical developments can be given here, but the following remarks may convey some intuitive understanding of the various approaches and how they hang together. The reader is referred to the literature cited in "Suggested Readings" (p. 281ff.) for further study.

It is generally agreed that "system" is a *model* of general nature, that is, a conceptual analog of certain rather universal traits of observed entities. Using models or analog constructs is the general procedure of science (and even of everyday cognition), as it is also the principle of analog simulation by computer. The difference from conventional disciplines is not essential but rather in the degree of generality (or abstraction), "system" refers to very general characteristics partaken by a large class of entities conventionally treated in different disciplines. Hence the interdisciplinary nature of general system theory; at the same time, its statements pertain to formal or structural communalities abstracting from the "nature of elements and forces in the system" with which the special sciences (and explanations in these) are concerned. In other words, system-theoretical arguments pertain to, and have predictive value inasmuch as such general structures are concerned. Such "explanation in principle" (pp. 36, 47 and elsewhere), may have considerable predictive value; for specific explanation, introduction of the special system conditions is naturally required (cf. p. 84ff.).

As discussed previously, "system" is a new "paradigm" in science in comparison to the elementalistic approach and conceptions predominating in scientific thinking. It is therefore not surprising that mathematical system theory has been developed in a variety of approaches with different emphasis, focus of interest, mathematical techniques, etc. Moreover, these approaches elucidate different aspects, properties, principles of what is comprised under the term, system, and serve different purposes of theoretical or practical nature. The fact that "system theories" by various authors look rather different is therefore not an embarrassment or result of confusion, but a healthy development in a new and growing field, and indicates presumably necessary and complementary aspects of the problem. The existence of different descriptions is nothing particular but is often encountered in mathematics and science, from the geometrical or analytical description of a curve to the equivalence of classical thermodynamics and statistical mechanics to that of wave mechanics and particle physics. Different and partly opposing approaches should, however, tend towards further integration, in the sense that one is a special case within another, or that they can be shown to be equivalent or complementary. Such developments are, in fact, taking place in system theory.

A system may be defined as a set of elements standing in interrelation among themselves and with environment. This can be mathematically expressed in different ways. Several typical ways of system description can be indicated.

One approach or group of investigations may, somewhat loosely, be circumscribed as *axiomatic* inasmuch as the focus of interest is in a rigorous definition of system and derivation, by modern methods of mathematics and logic, of its implications. Among others, examples are the the system descriptions by Mesarovic (1961 ff.), Maccia (1966), Beier and Laue (1971; set theory), Ashby (1958; "state-determined system or machine") Klir (1969) (UC = set of all couplings between the elements and the elements and environment; ST = set of all states and all transitions between states), etc.

Dynamical system theory is concerned with the changes of systems in time. There are two principal ways of description, i.e., internal and external description (cf. Rosen, 1971).

Internal description or "classical" system theory (Rosen, 1970) defines a system by a set of n measures, called state variables. Analytically, their change in time is typically expressed by a set of n simultaneous first-order differential equation (eq. 3.1, p. 54), called dynamical equations or equations of motion of the system. The behaviour of the system is described by the theory of differential equations (ordinary, first-order, if the definition of the system by eq. 3.1 is accepted) which is a well-known and highly developed field of mathematics. However, as was mentioned previously (p. 80), the system consideration poses quite definite problems. Thus, for example, the theory of stability has developed only recently in conjunction with problems of control (and system): the Liapunov († 1918) functions date from 1892 (in Russian, 1907 in French), but their significance was recognized only recently, especially through the work of mathematicians of the U.S.S.R.

Geometrically, the change of the system is expressed by the trajectories the state variables traverse in the state space, i.e., the n = dimensional space of possible location of the state variables. Three types of behaviour may be distinguished and defined as follows:

(1) A trajectory is called *asymptotically stable* if all trajectories sufficiently close to it at $t = t_o$, approach it asymptotically when $t \to \infty$.

(2) A trajectory is called *neutrally stable* if all trajectories sufficiently close to it at $t = 0$, remain close to it for all later time but do not necessarily approach it asymptotically.

(3) A trajectory is called *unstable* if the trajectories close to it at $t = 0$, do not remain close to it as $t \to \infty$.

This corresponds to solutions approaching a time-independent state (equilibrium, steady state), periodic solutions, and divergent solutions, respectively.

A time-independent state:

$$f_i (Q_1, Q_2, \ldots Q_n) = 0$$

can be considered as a trajectory degenerated into a single point. Then, readily visualizable in two-dimensional projection, the trajectories may converge towards a stable node represented by the equilibrium point, may approach it as stable focus in damped oscillations, or cycle around it in undamped oscillations (stable solutions); or else diverge from an unstable node, travel away from an unstable focus in oscillations, or from a saddle point (unstable solutions).

A central notion of dynamical theory is that of *stability*, i.e. the response of a system to perturbation. The concept of stability originates in mechanics (a rigid body is in stable equilibrium if it returns to its original position after sufficiently small displacement; a motion is stable if insensitive to small perturbations), and is generalized to the "motions" of the state variables of a system. This question is related to that of the existence of equilibrium states. Stability therefore can be analysed by explicit solution of the differential equations 3.1 describing the system (so-called indirect method, essentially based on discussion of the eigenwerte λ_i of the system of equations). In the case of non-linear systems, these have to be linearized by development in Taylor series and retaining the first term. Linearization, however, pertains only to stability in the vicinity of equilibrium. But stability arguments without actual solution of the differential equations (direct method) and for non-linear systems are possible by introduction of so-called *Liapunov functions* which are essentially generalized energy functions, the sign of which indicates whether or not an equilibrium is asymptotically stable (cf. La Salle and Lefschetz, 1961; Hahn, 1963).

Here the relation of dynamical system theory to control theory becomes apparent: control essentially means that a system which is not asymptotically stable, is made so by incorporating a controller counteracting motion of the system away from the stable state. For this reason the theory of stability in internal description or dynamical system theory converges with the theory of (linear) control or feedback systems in external description (cf. Schwarz, 1969).

Description by ordinary differential equations (eq. 3.1) abstracts from variations of the state variables in space which would be expressed by partial differential equations. These (field equations) are however more difficult to handle. Ways of overcoming this difficulty are the assumption of complete "stirring" so that distribution is homogenous within the volume considered; or the assumption of compartments to which homogenous distribution applies, and which are connected by suitable interactions (*compartment theory*).

In *external description,* the system is considered as a "black box"; its relations to environment and other systems are graphically presented in block and flow diagrams. The system description is in terms of inputs and outputs (*"Klemmenverhalten"* in German terminology): its general form are transfer functions relating input and output. Typically, these are assumed to be linear and are represented by discrete sets of values (cf. yes = no decisions in information theory, Turing machine). This is the language of control technology; external description, typically, is in terms of communication (exchange of information between system and environment and within the system) and control of the system's function with respect to environment (feedback), to use Wiener's definition of cybernetics.

As mentioned, internal and external descriptions largely coincide with descriptions by continuous or discrete functions. These are two "languages" adapted to their respective purposes. Empirically, there is—as emphasized in the text—an obvious contrast between regulations owing to the free interplay of forces within a dynamical system, and regulations owing to imposed constraints by structural feedback mechanisms. Formally, however, the two "languages" are related and in certain cases demonstrably translatable. For example, an input–output function can (under certain conditions) be developed as a linear n^{th}—order differential equation, and the terms of the latter considered as—formal—"state variables"; while their physical meaning remains indefinite, formal "translation" from one language into the other is thus possible.

In certain cases—for example the two-factor theory of nerve excitation (in terms of "excitatory and inhibitory factors" or

"substances") and network theory (McCulloch nets of "neurons")
—description in dynamical system theory by continuous functions
and description in automata theory by digital analogs can be
shown to be equivalent (Rosen, 1967). Similarly, predator–prey
systems, usually described dynamically by Volterra equations, can
also be expressed in terms by cybernetic feedback circuits (Wilbert, 1970). These are two-variables systems. Whether a similar
"translation" can be effectuated in many-variables systems remains (in the present writer's opinion) to be seen.

Internal description is essentially "structural", i.e. trying to
describe the system's behaviour in terms of state variables and
their interdependence. External description is "functional", describing the system's behaviour by its interaction with environment.

Thus, as this sketchy survey shows, considerable progress was
made in mathematical system theory since the program was
enunciated and commenced some twenty-five years ago, leading to
a variety of approaches which, however, are connected with each
other.

Mathematical system theory is a rapidly growing field, but it is
natural that basic problems such as, e.g., those of hierarchical
order (cf. p. 25ff.; Whyte, Wilson and Wilson, 1969) are only
slowly approached and presumably will need novel ideas and
theories. As we have stated (p. 36), general system theory ultimately is a "logico-mathematical science of wholeness" and its
rigorous development is "technical" = mathematical, but "verbal"
descriptions and models (as, for example, in Miller, 1969, Koestler,
1971, Weiss, 1970, Buckley, 1968, Gray, Duhl and Rizzo, 1968,
Demerath and Peterson, 1967 and elsewhere; cf. p. 22f.) are not
expendable. Problems must be intuitively "seen" and recognized
before they can be formalized mathematically. Otherwise, mathematical formalism may rather impede exploration of very "real"
problems.

Appendix II: The Meaning and Unity of Science*

At a time of universal crises such as we are experiencing today, the question of the meaning and purpose of natural sciences arises. That science is to be blamed for the miseries of our time is a reproach frequently heard; it is believed that men have been enslaved by machines, by technology at large, and eventually have been driven into the carnage of the world wars. We do not have the power to substantially influence the course of history; our only choice is to recognize it or to be overrun by it.

A renowned scholar, Professor Dr. Ludwig von Bertalanffy, addressed a crowded audience in the Department of Forensic Medicine within the framework of a scientific lecture series sponsored by FÖST (Freie Österreichische Studentenschaft). He spoke on vital present-day questions in connection with the problem of the special position of man in nature.

In contrast to the animal which has an "ambient" (*Umwelt*) determined by its organization, man himself creates his world, which we call human culture. Among the presuppositions for its evolution are two factors, language and formation of concepts, which are closely related to each other. "Language" as appeal or command can already be observed in the animal world; examples for this are the singing of birds, the warning whistle of mountain chamois, etc. Language as representation and communica-

*Review of a lecture at the University of Vienna, 1947.

tion of facts, however, is man's monopoly. Language, in the wider sense of the word, comprises not only oral speech but also script and the symbolic system of mathematics. These are systems not of inherited but of freely created and traditional *symbols*. First of all, this explains the *specificity of human history* in contrast to biological evolution: Tradition in contrast to hereditary mutations which occur only over a long period of time. Secondly, physical trial-and-error, largely characteristic of animal behavior, is replaced by *mental experimentation*—i.e., one with conceptual symbols. For this reason, true *goal-directedness* becomes possible. Goal-directedness and teleology in a metaphorical sense—i.e., regulation of happenings in the sense of maintenance, production and reproduction of organic wholeness, is a general criterion of life. True purposiveness, however, implies that actions are carried out with knowledge of their goal, of their future final results; the conception of the future goal does already exist and influences present actions. This applies to primitive actions of everyday life as well as to the highest achievements of the human intellect in science and technology. Furthermore, the symbolic world created by men gains a life of its own, as it were; it becomes more intelligent than its creator. The symbol system of mathematics, for example, is embodied in an enormous thinking machine which, fed with a statement, produces in return a solution on the basis of a fixed process of concatenation of symbols, which could hardly be anticipated. On the other hand, however, this symbolic world becomes a power which can lead to grave disturbances. If it comes to a conflict between the symbolic world— which in human society has emerged in moral values and social conventions—and biological drives, which are out of place in cultural surroundings, the individual is confronted with a situation prone to psychoneurosis. As a social power the symbolic world, which makes man human, at the same time produces the sanguinary course of history. In contrast to the naive struggle for existence of organisms, human history is largely dominated by the struggle of ideologies—i.e., of symbolisms, which are the more dangerous, the more they disguise primitive instincts. We cannot revoke the course of events, which has produced what we call "man"; it is up to him, however, whether he applies his power of foresight for his enhancement or for his own annihilation. In this sense the question of what course the scientific world-conception will take is at the same time a question of the destiny of mankind.

A survey of scientific developments reveals a strange phenomenon. Independently of each other, *similar general principles* start to take shape in the various fields of science. As such, the lecturer emphasized especially the aspects of organization, wholeness, and dynamics, and sketched their influence in the various sciences. In physics, these conceptions are characteristic of modern in contrast to classical physics. In biology, they are emphasized by the *"organismic conception"* represented by the lecturer. Similar conceptions are found in medicine, psychology (gestalt psychology, theory of stratification) and in modern philosophy.

This results in a tremendous perspective, the prospect of a unity of the world-view hitherto unknown. How does this unity of general principles come about? Dr. von Bertalanffy answers this question by demanding a *new field in science* which he calls "General System Theory" and which he attempted to found. This is a logico-mathematical field, whose task is the formulation and derivation of those general principles that are applicable to "systems" in general. In this way, exact formulations of terms such as wholeness and sum, differentiation, progressive mechanization, centralization, hierarchical order, finality and equifinality, etc., become possible, terms which occur in all sciences dealing with "systems" and imply their logical homology.

Last century's mechanistic world picture was closely related to the domination of the machine, the theoretical view of living beings as machines and the mechanization of man himself. Concepts, however, which are coined by modern scientific developments, have their most obvious exemplification in life itself. Thus, there is a hope that the new world concept of science is an expression of the development toward a new stage in human culture.

REFERENCES

ACKOFF, R. L., "Games, Decisions and Organization," *General Systems*, 4 (1959), 145–150.

———, "Systems, Organizations, and Interdisciplinary Research," *General Systems*, 5 (1960), 1–8.

ADAMS, H., *The Degradation of the Democratic Dogma*, New York, Macmillan, 1920.

ADOLPH, E. F., "Quantitative Relations in the Physiological Constitution of Mammals," *Science*, 109 (1949), 579–585.

AFANASJEW, W. G., "Über Bertalanffy's 'organismische' Konzeption," *Deutsche Zeitschrift für Philosophie*, 10 (1962), 1033–1046.

ALEXANDER, FRANZ, *The Western Mind in Transition: An Eyewitness Story*, New York, Random House, 1960.

ALLESCH, G. J. von, *Zur Nichteuklidischen Struktur des phænomenalen Raumes*, Jena, Fischer, 1931.

ALLPORT, FLOYD, *Theories of Perception and the Concept of Structure*, New York, John Wiley & Sons, 1955.

ALLPORT, GORDON W., *Becoming: Basic Considerations for a Psychology of Personality*, New Haven, Yale University Press, 1955.

———, "European and American Theories of Personality, *Perspectives in Personality Theory*, Henry David and Helmut von Bracken, editors, London, Tavistock, 1957.

———, "The Open System in Personality Theory," *Journal of Abnormal and Social Psychology*, 61 (1960), 301–310. (Reprinted in *Personality and Social Encounter*, Boston, Beacon Press, 1960).

———, *Pattern and Growth in Personality*, New York, Holt, Rinehart & Winston, 1961.

ANDERSON, HAROLD, "Personality Growth: Conceptual Considerations," *Perspectives in Personality Theory*, Henry David and Helmut von Bracken, editors, London, Tavistock, 1957.

ANON., "Crime and Criminology," *The Sciences*, 2 (1963), 1–4.

ANSCHÜTZ, G., *Psychologie*, Hamburg, Meiner, 1953.

APPLEBY, LAWRENCE, J. SCHER and J. CUMMINGS, editors, *Chronic Schizophrenia*, Glencoe, Ill., The Free Press, 1960.

ARIETI, SILVANO, *Interpretation of Schizophrenia*, New York, Robert Brunner, 1955.

———, "Schizophrenia," *American Handbook of Psychiatry*, S. Arieti, editor, vol. 1, New York, Basic Books, 1959.

———, "The Microgeny of Thought and Perception," *Arch. of Gen. Psychiat.* 6 (1962), 454–468.

———, "Contributions to Cognition from Psychoanalytic Theory," *Science and Psychoanalysis*, G. Masserman (ed.), Vol. 8, New York, Grune & Strutton, 1965.

ARROW, K. J., "Mathematical Models in the Social Sciences," *General Systems*, I (1956), 29–47.

ASHBY, W. R., "Can a Mechanical Chess-Player Outplay its Designer?" *British Journal of Philos. Science*, 3 (1952a), 44.

———, *Design for a Brain*, London, Chapman & Hall, 1952b.

———, "General Systems Theory as a New Discipline," *General Systems*, 3 (1958a), 1–6.

———, *An Introduction to Cybernetics*, 3rd. edition, New York, John Wiley & Sons, 1958b.

———, "Principles of the Self-Organizing System," *Principles of Self-Organization*, H. von Foerster and G. W. Zopf, Jr., editors, New York, Pergamon Press, 1962.

———, "Constraint Analysis of Many-Dimensional Relations," *Technical Report #2*, May 1964, Urbana, Electrical Engineering Research Laboratory, University of Illinois.

ATTNEAVE, F., *Application of Information Theory to Psychology*, New York, Holt, Rinehart & Winston, 1959.

BACKMAN, G., "Lebensdauer und Entwicklung," *Roux' Arch.*, 140 (1940), 90.

BAVINK, B., *Ergebnisse und Probleme der Naturwissenschaften*, 8th ed., Leipzig, Hirzel, 1944; 9th ed., Zurich, Hirzel, 1949.

BAYLISS, L. E., *Living Control Systems*, San Francisco, Freeman, 1966.

BEADLE, G. W., *Genetics and Modern Biology*, Philadelphia, American Philosophical Society, 1963.

BECKNER, M., *The Biological Way of Thought*, New York, Columbia University Press, 1959.

BEER, S., "Below the Twilight Arch—A Mythology of Systems," *General Systems*, 5 (1960), 9–20.

BEIER, W., *Biophysik*, 2. Aufl., Leipzig, Thieme, 1962.

———, *Einführung in die theoretische Biophysik*, Stuttgart, Gustav Fischer, 1965.

BELL, E., "Oogenesis," C. P. Raven, review, *Science*, 135 (1962), 1056.

BENDMANN, A., "Die 'organismische Auffassung' Bertalanffys," *Deutsche Zeitschrift für Philosophie*, 11 (1963), 216–222.

———, *L. von Bertalanffys organismische Auffassung des Lebens in ihren philosophischen Konsequenzen*, Jena, G. Fischer, 1967.

BENEDICT, RUTH, *Patterns of Culture*, New York, (1934) Mentor Books, 1946.

BENTLEY, A. F., "Kenetic Inquiry," *Science*, 112 (1950), 775.

BERG, K., and K. W. OCKELMANN, "The Respiration of Freshwater Snails," *J. Exp. Biol.*, 36 (1959), 690–708.

———, "On the Oxygen Consumption of Some Freshwater Snails," *Verh. Int. Ver. Limnol.*, 14 (1961), 1019–1022.

BERLIN, SIR ISAIAH, *Historical Inevitability*, London, New York, Oxford University Press, 1955.

BERLYNE, D. E., "Recent Developments in Piaget's Work," *Brit. J. of Educ. Psychol.* 27 (1957), 1–12.

———, *Conflict, Arousal and Curiosity*, New York, McGraw-Hill, 1960.

BERNAL, J. D., *Science in History*, London, Watts, 1957.

BERTALANFFY, F. D., and C. LAU, "Cell Renewal," *Int. Rev. Cytol.*, 13 (1962), 357–366.

BERTALANFFY, LUDWIG von, "Einführung in Spengler's Werk," *Literaturblatt Kölnische Zeitung* (May, 1924).

————, *Kritische Theorie der Formbildung*, Berlin, Borntraeger, 1928a. English: *Modern Theories of Development* (1934), New York, Harper Torchbooks, 1962.

————, *Nikolaus von Kues*, Munich, G. Müller, 1928b.

————, *Theoretische Biologie*, Bd. I, II, Berlin, Borntraeger, 1932, 1942 (2nd ed. Bern, A. Francke AG., 1951).

"Untersuchungen über die Gesetzlichkeit des Wachstums, I." *Roux' Archiv*, 131 (1934), 613–652.

————, *Das Gefüge des Lebens*, Leipzig, Teubner, 1937.

————, "II. A Quantitative Theory of Organic Growth," *Human Biology* 10 (1938), 181–213.

————, "Der Organismus als physikalisches System betrachtet," *Die Naturwissenschaften*, 28 (1940a), 521–531. Chapter 5.

————, "Untersuchungen über die Gesetzlichkeit des Wachstums," III, Quantitative Beziehungen zwischen Darmoberflache und Körpergrösse bei *Planaria Maculata*," *Roux' Arch.*, 140 (1940b), 81–89.

————, "Probleme einer dynamischen Morphologie," *Biologia Generalis*, 15 (1941), 1–22.

————, "Das Weltbild der Biologie"; "Arbeitskreis Biologie," *Weltbild und Menschenbild*, S. Moser, editor (Alpbacher Hochshulwochen 1947), Salzburg, Tyrolia Verlag, 1948a.

————, "Das organische Wachstum und seine Gesetzmässigkeiten," *Experientia* (Basel), 4 (1948b).

————, *Das biologische Weltbild*, Bern, A. Francke AG, 1949a. English: *Problems of Life*, New York, Wiley, 1952; New York, Harper Torchbooks, 1960. Also in French, Spanish, Dutch, Japanese.

————, "Problems of Organic Growth," *Nature*, 163 (1949b), 156.

————, "Zu einer allgemeinen Systemlehre," *Blätter für deutsche Philosophie*, 3/4 (1945), (Extract in *Biologia Generalis*, 19 (1949), 114–129).

————, "The Theory of Open Systems in Physics and Biology," *Science*, 111 (1950a), 23–29.

————, "An Outline of General System Theory," *Brit. J. Philos. Sci.* 1 (1950b), 139–164.

————, "Theoretical Models in Biology and Psychology," *Theoretical Models and Personality Theory*, D. Krech and G. S. Klein, editors, Durham, Duke University Press, 1952.

————, *Biophysik des Fliessgleichgewichts*, translated by W. H. Westphal, Braunschweig, Vieweg, 1953a. (Revised edition with W. Beier and R. Laue in preparation).

————, "Philosophy of Science in Scientific Education," *Scient. Monthly*, 77 (1953b), 233.

————, "General System Theory." *Main Currents in Modern Thought*, 11 (1955a), 75–83. Chapter 2.

————, "An Essay on the Relativity of Categories," *Philosophy of Science*, 22 (1955b), 243–263. Chapter 10.

————, "A Biologist Looks at Human Nature," *Scientific Monthly*, 82 (1956a), 33–41. Reprinted in *Contemporary Readings in Psychology*, R. S. Daniel, editor, 2nd edition, Boston, Houghton Mifflin Company, 1965. Also in *Reflexes to Intelligence, A Reader in Clinical Psychology*, S. J. Beck and H. B. Molish, editors, New York, Glencoe (Ill.): The Free Press, 1959.

————, "Some Considerations on Growth in its Physical and Mental Aspects," *Merrill-Palmer Quarterly*, 3 (1956b), 13–23.

————, *The Significance of Psychotropic Drugs for a Theory of Psychosis*, World Health Organization, AHP, 2, 1957a. (Mimeograph).

————, "Wachstum," *Kükenthal's Handb. d. Zoologie*, Bd. 8, 4 (6), Berlin: De Gruyter, 1957b.

————, "Comments on Aggression," *Bulletin of the Menninger Clinic*, 22 (1958), 50–57.

————, "Human Values in a Changing World," *New Knowledge in Human Values*, A. H. Maslow, editor, New York, Harper & Brothers, 1959.

————, "Some Biological Considerations on the Problem of Mental Illness," *Chronic Schizophrenia*, L. Appleby, J. Scher, and J. Cummings, editors, Glencoe (Ill.): The Free Press, 1960a. Reprinted in *Bulletin of the Menninger Clinic*, 23 (1959), 41–51.

————, "Principles and Theory of Growth," *Fundamental Aspects of Normal and Malignant Growth*, W. W. Nowinski, editor, Amsterdam, Elsevier, 1960b.

————, "General System Theory—A Critical Review," *General Systems*, 7 (1962), 1–20. Chapter 4.

————, "The Mind-Body Problem: A New View," *Psychosom. Med.*, 24 (1964a) 29–45.

————, "Basic Concepts in Quantitative Biology of Metabolism," *Quantitative Biology of Metabolism—First International Symposium*, A. Locker, editor. *Helgoländer wissenschaftliche Meeresuntersuchungen*, 9 (1964b), 5–37.

————, "The World of Science and the World of Value," *Teachers College Record*, 65 (1964c), 496–507.

————, "On the Definition of the Symbol," *Psychology and the Symbol: An Interdisciplinary Symposium*, J. R. Royce, editor, New York, Random House, 1965.

————, "General System Theory and Psychiatry," *American Handbook of Psychiatry*, vol. 3, S. Arieti, editor, New York, Basic Books, 1966. Chapter 9.

————, *Robots, Men and Minds*, New York, George Braziller, 1967.

————, and R. R. ESTWICK, "Tissue Respiration of Musculature in Relation to Body Size," *Amer. J. Physiol.* 173 (1953), 58–60.

————, C. G. HEMPEL, R. E. BASS and H. JONAS, "General System Theory: A New Approach to Unity of Science," I-VI, *Hum. Biol.* 23 (1951), 302–361.

————, and I. MÜLLER, "Untersuchungen über die Gesetzlichke des Wachstums, VIII, Die Abhängigkeit des Stoffwechsels von d Körpergrösse und der Zusammenhang von Stoffwechseltypen ur Wachstumstypen," *Riv. Biol.*, 35 (1943), 48–95.

————, and W. J. P. PIROZYNSKI, "Ontogenetic and Evolutiona Allometry," *Evolution*, 6 (1952), 387–392.

————, "Tissue Respiration, Growth and Basal Metabolism," *Bio Bull.*, 105 (1953), 240–256.

BETHE, ALBERT, "Plastizität und Zentrenlehre," *Handbuch d normalen und pathologischen Physiologie*, vol. XV/2, Albert Beth editor, Berlin, Springer, 1931.

BEVERTON, R. J. H., and S. J. Holt, "On the Dynamics of E ploited Fish Populations," *Fishery Investigation*, Ser. II, vol. XI London, Her Majesty's Stationery Office, 1957.

BLANDINO, G., *Problemi e Dottrine di Biologia teorica*, Bologn Minerva Medica, 1960.

BLASIUS, W., "Erkenntnistheoretische und methodologische Grun lagen der Physiologie," in Landois-Rosemann, *Lehrbuch der Phy iologie des Menschen*, 28. Aufl., Munich, Berlin, Urban & Schwarze berg, 1962, 990–1011.

BLEULER, EUGEN, *Mechanismus-Vitalismus-Mnemismus*, Berli Springer, 1931.

BODE, H., F. MOSTELLER, F. TUKEY, and C. WINSOR, "Th Education of a Scientific Generalist," *Science*, 109 (1949), 553.

BOFFEY, PHILIP M., "Systems Analysis: No Panacea for Nation Domestic Problems," *Science*, 158 (1967), 1028–1030.

BOGUSLAW, W., *The New Utopians*, Englewood Cliffs, Prentice-Hal 1965.

BOULDING, K. E., *The Organizational Revolution*, New York, Harpe & Row, 1953.

————, "Toward a General Theory of Growth," *General Systems*, (1956a), 66–75.

————, *The Image*, Ann Arbor, University of Michigan Press, 1956l

————, *Conflict and Defence*. New York: Harper, 1962.

BRADLEY, D. F., and M. CALVIN, "Behavior: Imbalance in a Ne work of Chemical Transformation," *General Systems*, I (1956 56–65.

BRAY, H. G., and K. WHITE, *Kinetics and Thermodynamics in Bio chemistry*, New York, Academic Press, 1957.

BRAY, J. R., "Notes toward an Ecology Theory," *Ecology*, 9 (1958 770–776.

BRODY, S., "Relativity of Physiological Time and Physiologica Weight," *Growth*, 1 (60), 1937.

————, *Bioenergetics and Growth*, New York, Reinhold, 1945.

BRONOWSKI, J., Review of "Brains, Machines and Mathematics" by M. A. Arbib, *Scientific American*, (July 1964), 130–134.

BRUNER, JEROME, "Neural Mechanisms in Perception," in *The Brain and Human Behavior*, H. Solomon, editor, Baltimore, Williams and Wilkins, 1958.

BRUNNER, R., "Das Fliessgleichgewicht als Lebensprinzip," *Mitteilungen der Versuchsstation für das Gärungsgewerbe* (Vienna), 3–4 (1967), 31–35.

BRUNSWIK, EGON, "Historical and Thematic Relations of Psychology to Other Sciences," *Scientific Monthly*, 83 (1956), 151–161.

BUCKLEY, W., *Sociology and Modern Systems Theory*, Englewood Cliffs, N. J., Prentice-Hall, 1967.

BÜHLER, C., "Theoretical Observations about Life's Basic Tendencies," *Amer. J. Psychother.*, 13 (1959), 561–581.

————, *Psychologie im Leben unserer Zeit*, Munich & Zurich, Knaur, 1962.

BURTON, A. C., "The Properties of the Steady State Compared to those of Equilibrium as shown in Characteristic Biological Behavior," *J. Cell. Comp. Physiol.*, 14 (1939), 327–349.

BUTENANDT, A., "Neuartige Probleme und Ergebnisse der biologischen Chemie," *Die Naturwissenschaften*, 42 (1955), 141–149.

————, "Altern und Tod als biochemisches Problem," *Dt. Med. Wschr.*, 84 (1959), 297–300.

CANNON, W. B., "Organization for Physiological Homeostasis," *Physiological Review*, 9 (1929), 397.

————, *The Wisdom of the Body*, New York, W. W. Norton Co., 1932.

CANTRIL, HADLEY, "A Transaction Inquiry Concerning Mind," *Theories of the Mind*, Jordan Scher, editor, New York, The Free Press, 1962.

CARMICHAEL, LEONARD, editor, *Manual of Child Psychology* (2nd edition), New York, John Wiley & Sons, 1954.

CARNAP, R., *The Unity of Science*, London, 1934.

CARTER, L. J., "Systems Approach: Political Interest Rises," *Science*, 153 (1966), 1222–1224.

CASEY, E. J., *Biophysics*, New York, Reinhold, 1962.

CASSIRER, ERNST, *The Philosophy of Symbolic Forms*, 3 vols., New Haven, Yale University Press, 1953–1957.

CHANCE, B., R. W. ESTABROOK & J. R. WILLIAMSON (eds.), *Control of Energy Metabolism*, New York, London, Academic Press, 1965.

CHOMSKY, N., " 'Verbal Behavior' by B. F. Skinner," *Language*, 35 (1959), 26–58.

CHORLEY, R. J., "Geomorphology and General Systems Theory," *General Systems*, 9 (1964), 45–56.

COMMONER, B., "In Defense of Biology," *Science*, 133 (1961), 1745–1748.

COWDRY, EDMUND, *Cancer Cells*, 2nd edition, Philadelphia, W. B. Saunders, 1955.

(DAMUDE, E.), "A Revolution in Psychiatry," *The Medical Post*, May 23, 1967.

D'ANCONA, V., *Der Kampf ums Dasein*, Berlin, Bornträger, 1939. English translation, *The Struggle for Existence*, Leiden, E. J. Brill, 1954.

DENBIGH, K. G., "Entropy Creation in Open Reaction Systems," *Trans. Faraday. Soc.*, 48 (1952), 389–394.

DE-SHALIT, A., "Remarks on Nuclear Structure," *Science*, 153 (1966), 1063–1067.

DOBZHANSKY, T., "Are Naturalists Old-Fashioned?" *American Naturalist*, 100 (1966), 541–550.

DONNAN, F. G., "Integral Analysis and the Phenomenon of Life," *Acta Biotheor*, 1937.

DOST, F. H., *Der Blutspiegel: Kinetik der Konzentrationsabläufe in der Körperflüssigkeit*, Leipzig, Thieme, 1953.

———, "Über ein einfaches statistisches Dosis-Umsatz-Gesetz," *Klin. Wschr.*, 36 (1958), 655–657.

———, "Beitrag zur Lehre vom Fliessgleichgewicht (steady state) aus der Sicht der experimentellen Medizin," *Nova Acta Leopoldina*, 4–5 (1958–59), 143–152.

———, "Fliessgleichgewichte im strömenden Blut," *Dt. med. Wschr.*, 87 (1962a), 1833–1840.

———, "Ein Verfahren zur Ermittlung des absoluten Transportvermögens des Blutes im Fliessgleichgewicht, *Klin. Wschr.*, 40 (1962b), 732–733.

DRISCHEL, H., "Formale Theorien der Organisation (Kybernetik und verwandte Disziplinen)," *Nova Acta Leopoldina* (Halle, Germany), (1968).

DRUCKREY, H., and K. KUPFMÜLLER, *Dosis und Wirkung. Die Pharmazie*, 8. Beiheft, 1, Erg.-Bd., Aulendorf (Württ.) Editio Cantor GmbH., 1949, 513–595.

DUBOS, R., "Environmental Biology," *BioScience*, 14 (1964), 11–14.

———, "We are Slaves to Fashion in Research!", *Scientific Research* (Jan. 1967), 36–37, 54.

DUNN, M. S., E. A. MURPHY and L. B. ROCKLAND, "Optimal Growth of the Rat," *Physiol. Rev.*, 27 (1947), 72–94.

EGLER, F. E., "Bertalanffian Organismicism," *Ecology*, 34 (1953), 443–446.

ELSASSER, W. M., *The Physical Foundation of Biology*, New York, Pergamon Press, 1958.

———, *Atom and Organism*, Princeton, Princeton University Press, 1966.

EYSENCK, HANS, "Characterology, Stratification Theory and Psychoanalysis; An Evaluation," *Perspectives in Personality Theory*,

Henry David and Helmut von Bracken, editors, London, Tavistock, 1957.

FEARING, F., "An Examination of the Conceptions of Benjamin Whorf in the Light of Theories of Perception and Cognition," *Language in Culture*, H. Hoijer, editor, *American Anthropologist*, 56 (1954), Memoir No. 79, 47.

FLANNERY, Kent V., "Culture History v. Cultural Process: A Debate in American Archaeology," *Sci. Amer.*, 217 (1967), 119–122.

FOERSTER, H. von and G. W. ZOPF, Jr. (eds.), *Principles of Self-Organization*, New York, Pergamon Press, 1962.

FOSTER, C., A. RAPOPORT and E. TRUCCO, "Some Unsolved Problems in the Theory of Non-Isolated Systems," *General Systems*, 2 (1957), 9–29.

FRANK, L. K., G. E. HUTCHINSON, W. K. LIVINGSTONE, W. S. McCULLOCH, and N. WIENER, *Teleological Mechanisms*, Ann. N.Y. Acad. Sci., 50 (1948).

FRANKL, VICTOR, "Das homöostatische Prinzip und die dynamische Psychologie," *Zeitschrift für Psychotherapie und Medizinische Psychologie*, 9 (1959a), 41–47.

——, *From Death-Camp to Existentialism*, Boston, Beacon Press, 1959b.

——, "Irrwege seelenärztlichen Denkens (Monadologismus, Potentialismus und Kaleidoskopismus)," *Nervenarzt*, 31 (1960), 385–392.

FRANKS, R. G. E., *Mathematical Modeling in Chemical Engineering*, New York, Wiley, 1967.

FREEMAN, GRAYDON, *The Energetics of Human Behavior*, Ithaca, Cornell University Press, 1948.

FREUD, SIGMUND, *A General Introduction to Psychoanalysis*, (1920) New York, Permabooks, 1953.

FRIEDELL, E., *Kulturgeschichte der Neuzeit*, München, C. H. Beck, 1927–31.

GARAVAGLIA, C., C. POLVANI and R. SILVESTRINI, "A Collection of Curves Obtained With a Hydrodynamic Model Simulating Some Schemes of Biological Experiments Carried Out With Tracers," Milano, CISE, *Report No. 60* (1958), 45 pp.

GAZIS, DENOS C., "Mathematical Theory of Automobile Traffic," *Science*, 157 (1967), 273–281.

GEERTZ, CLIFFORD, "The Growth of Culture and the Evolution of Mind," *Theories of the Mind*, Jordan Scher, editor, New York, The Free Press, 1962.

GESSNER, F., "Wieviel Tiere bevölkern die Erde?" *Orion* (1952), 33–35.

GEYL, P., *Napoleon For and Against*, London, Jonathan Cape, 1949. (1957)

——, *Debates With Historians*, New York, Meridian Books, 1958.

GILBERT, ALBIN, "On the Stratification of Personality," *Perspectives in Personality Theory*, Henry David and Helmut von Bracken, editors, London, Tavistock, 1957.

GILBERT, E. N., "Information Theory After 18 Years," *Science*, 152 (1966), 320–326.

GLANSDORFF, P. and J. PRIGOGINE, "On a General Evolution Criterion in Macroscopic Physics," *Physica* 30 (1964), 351–374.

GOLDSTEIN, KURT, *The Organism*, New York, American Book Company, 1939.

————, "Functional Disturbances in Brain Damage," *American Handbook of Psychiatry*, vol. 1, Silvano Arieti, editor, New York, Basic Books, 1959.

GRAY, W., N. D. RIZZO and F. D. DUHL (eds.), *General Systems Theory and Psychiatry*, Boston, Little, Brown and Company, (in press).

GRINKER, R. R. (ed.), *Toward a Unified Theory of Human Behavior*, 2nd edition, New York, Basic Books, 1967.

GRODIN, F. S., *Control Theory and Biological Systems*, New York, Columbia University Press, 1963.

GROSS, J., "Die Krisis in der theoretischen Physik und ihre Bedeutung für die Biologie," *Biologisches Zentralblatt*, 50 (1930).

GUERRA, E. and B. GÜNTHER, "On the Relationship of Organ Weight, Function and Body Weight," *Acta Physiol. Lat. Am.*, 7 (1957), 1–7.

GÜNTHER, B. and E. GUERRA, "Biological Similarities," *Acta Physiol. Lat. Am.*, 5 (1955), 169–186.

HACKER, FREDERICK, "Juvenile Delinquency," *Hearings before the U. S. Senate Subcommittee Pursuant to S. Res. No. 62*, Washington, U.S. Government Printing Office, June 15–18, 1955.

HAHN, ERICH, "Aktuelle Entwicklungstendenzen der soziologischen Theorie," *Deutsche Z. Philos.*, 15 (1967), 178–191.

HAIRE, M., "Biological Models and Empirical Histories of the Growth of Organizations," *Modern Organization Theory*, M. Haire, editor, New York, John Wiley & Sons, 1959, pp. 272–306.

HALL, A. D., and R. E. FAGEN, "Definition of System," *General Systems*, I (1956), 18–29.

————, *A Methodology for Systems Engineering*, Princeton, Van Nostrand, 1962.

HALL, C. S., and G. LINDZEY, *Theories of Personality*, New York, John Wiley & Sons, 1957.

HART, H., "Social Theory and Social Change," in L. Gross, editor, *Symposium on Sociological Theory*, Evanston, Row, Peterson, 1959, pp. 196–238.

HARTMANN, M., *Allgemeine Biologie*, Jena, 1927.

HARTMANN, N., "Neue Wege der Ontologie," *Systematische Philosophie*, N. Hartmann, editor, Stuttgart, 1942.

HAYEK, F. A., "Degrees of Explanation," *Brit. J. Philos. Sci.*, 6 (1955), 209–225.

HEARN, G., *Theory Building in Social Work*, Toronto, University of Toronto Press, 1958.

HEBB, DONALD O., *The Organization of Behavior*, New York, John Wiley & Sons, 1949.

———, "Drives and the C.N.S. (Conceptual Nervous System)," *Psychol. Rev.*, 62 (1955), 243–254.

HECHT, S., "Die physikalische Chemie und die Physiologie des Sehaktes," *Erg. Physiol.*, 32 (1931).

HEMMINGSEN, A. M., "Energy Metabolism as Related to Body Size and Respiratory Surfaces, and its Evolution," *Reps. Steno Mem. Hosp.*, part 2, 9 (1960).

HEMPEL, C. G., *Aspects of Scientific Explanation and other Essays in the Philosophy of Science*, New York, The Free Press, 1965.

HENRY, JULES, *Culture Against Man*, New York, Random House, 1963.

HERRICK, CHARLES, *The Evolution of Human Nature*, New York, Harper Torchbooks, 1956.

HERSH, A. H., "Drosophila and the Course of Research," *Ohio J. Sci.*, 42 (1942), 198–200.

HESS, B., "Fliessgleichgewichte der Zellen," *Dt. Med. Wschr.*, 88 (1963), 668–676.

———, "Modelle enzymatischer Prozesse," *Nova Acta Leopoldina* (Halle, Germany), 1969.

HESS, B., and B. CHANCE, "Über zelluläre Regulationsmechanismen und ihr mathematisches Modell," *Die Naturwissenschaften*, 46 (1959), 248–257.

HESS, W. R., "Die Motorik als Organisationsproblem," *Biologisches Zentralblatt*, 61 (1941), 545–572.

———, "Biomotorik als Organisationsproblem," I, II, *Die Naturwissenschaften*, 30 (1942), 441–448, 537–541.

HILL, A. V., "Excitation and Accommodation in Nerve," *Proc. Roy. Soc. London*, 11 (1936).

HOAGLAND, H., "Consciousness and the Chemistry of Time," *Transactions of the First Conference*, H. A. Abramson, editor, New York, J. Macy Foundation, 1951.

HÖBER, R., *Physikalische Chemie der Zelle und der Gewebe*, 6. Aufl., Berlin, 1926.

HOIJER, H., editor, *Language in Culture*, American Anthropologist, Memoir No. 79, (1954).

HOLST, ERICH von, "Vom Wesen der Ordnung im Zentralnervensystem," *Die Naturwissenschaften*, 25 (1937), 625–631, 641–647.

HOLT, S. J., "The Application of Comparative Population Studies to Fisheries Biology—An Exploration," in *The Exploitation of Natural Animal Populations*, E. D. LeCren and M. W. Holdgate, editors, Oxford, Blackwell, n.d.

HOOK, SIDNEY, editor, *Dimensions of Mind*, New York, Collier Books, 1961.

HUMBOLDT, W. VON, *Gesammelte Schriften*, VII, 1, Berlin, Preuss. Akademie, n.d.

HUXLEY, A., *The Doors of Perception*, New York, Harper & Row, 1954.

HUXLEY, J., *Problems of Relative Growth*, London, Methuen, 1932.

JEFFRIES, L. A. (ed.), *Cerebral Mechanisms in Behavior*, The Hixon Symposium, New York, John Wiley & Sons, 1951.

JONES, R. W., and J. S. GRAY, System Theory and Physiological Processes," *Science*, 140 (1963), 461–466.

JUNG, F., "Zur Anwendung der Thermodynamik auf Biologische und Medizinische Probleme," *Die Naturwissenschaften*, 43 (1956), 73–78.

KALMUS, H., "Über die Natur des Zeitgedächtnisses der Bienen," *Zeitschr. Vergl. Physiol.*, 20 (1934), 405.

KAMARYT, J., "Die Bedeutung der Theorie des offenen Systems in der gegenwärtigen Biologie," *Deutsche Z. für Philosophie*, 9 (1961), 2040–2059.

————, "Ludwig von Bertalanffy a syntetické směry v západní biologii," in *Filosofické problémy moderni biologie*, J. Kamarýt, editor, Prague, Československá Akademie, VED, 1963, pp. 60–105.

KANAEV, I. I., *Aspects of the History of the Problem of the Morphological Type from Darwin to the Present* (in Russian), Moscow, NAUKA, 1966, pp. 193–200.

KEITER, F., "Wachstum und Reifen im Jugendalter," *Kölner Z. für Soziologie*, 4 (1951–1952), 165–174.

KLEIBER, M., *The Fire of Life*, New York, John Wiley & Sons, 1961.

KLUCKHOHN, C., and D. LEIGHTON, "The Navaho," *The Tongue of the People*, Cambridge (Mass.), Harvard University Press, 1951.

————, "Culture and Behavior," in *Handbook of Social Psychology*, vol. 2, G. Lindzey, editor, Cambridge, Addison-Wesley Publishing Company, 1954.

KMENT, H., "Das Problem biologischer Regelung und seine Geschichte in medizinischer Sicht," *Münch. Med. Wschr.*, 99 (1957), 475–478, 517–520.

————, "The Problem of Biological Regulation and Its Evolution in Medical View," *General Systems*, 4 (1959), 75–82.

KOESTLER, ARTHUR, *The Lotus and the Robot*, London, Hutchinson, 1960.

————, *The Ghost in the Machine*, London, Hutchinson, 1967.

————, "The Tree and the Candle," in *Unity and Diversity*, see *Addenda* (1971).

KÖHLER, W., *Die physischen Gestalten in Ruhe und im stationären Zustand*, Erlangen, 1924.

————, "Zum Problem der Regulation," *Roux's Arch*, 112 (1927).

KOTTJE, F., "Zum Problem der vitalen Energie," *Ann. d. Phil. u. phil. Kritik*, 6 (1927).

KRECH, DAVID, "Dynamic Systems as Open Neurological Systems," *Psychological Review*, 57 (1950), 283–290. Reprinted in *General Systems*, I (1956), 144–154.

KREMYANSKIY, V. I., "Certain Peculiarities of Organisms as a 'System' from the Point of View of Physics, Cybernetics, and Biology," *General Systems*, 5 (1960), 221–230.

KROEBER, A. L., *The Nature of Culture*, Chicago, The University of Chicago Press, 1952.

————, *Style and Civilizations*, Ithaca, N. Y., Cornell U. Press, 1957.

————, and C. KLUCKHOHN, *Culture. A Critical Review of Concepts and Definitions* (1952), New York, Vintage, 1963.

KUBIE, LAWRENCE, "The Distortion of the Symbolic Process in Neurosis and Psychosis," *J. Amer. Psychoanal. Ass.*, 1 (1953), 59–86.

KUHN, T. S., *The Structure of Scientific Revolutions*, Chicago, University of Chicago Press, 1962.

LABARRE, W., *The Human Animal*, Chicago, U. of Chicago Press, 1954.

LANGER, S., *Philosophy in a New Key* (1942), New York, Mentor Books, 1948.

LASHLEY, K., *Brain Mechanisms and Intelligence* (1929), New York, Hafner, 1964.

LECOMTE DU NOÜY, P., *Biological Time*, New York, Macmillan, 1937.

LEHMANN, G., "Das Gesetz der Stoffwechselreduktion," in *Kükenthals Handbuch der Zoologie*, 8, 4(5). Berlin, De Gruyter & Co., 1956.

LENNARD, H., and A. BERNSTEIN, *The Anatomy of Psychotherapy*, New York, Columbia U. Press, 1960.

LERSCH, P., and H. THOMAE, editors, *Handbuch der Psychologie*, Vol. 4: *Persönlichkeitsforschung und Persönlichkeitstheorie*. Göttingen, Hogrefe, 1960.

LEWADA, J., "Kybernetische Methoden in der Soziologie," (in Russian) *Kommunist* (Moscow, 1965), 14, 45.

LLAVERO, F., "Bemerkungen zu einigen Grundfragen der Psychiatrie," *Der Nervenarzt*, 28 (1957), 419–420.

LOCKER, A., "Das Problem der Abhängigkeit des Stoffwechsels von der Körpergrösse," *Die Naturwissenschaften*, 48 (1961a), 445–449.

————, "Die Bedeutung experimenteller Variabler für die Abhängigkeit der Gewebsatmung von der Körpergrösse. II. Die Bezugsbasis," *Pflügers Arch. ges. Physiol.*, 273 (1961b), 345–352.

————, "Reaktionen metabolisierender Systeme auf experimentelle Beeinflussung, Reiz und Schädigung," *Helgoländer wissenschaftliche Meeresuntersuchungen*, 9 (1964), 38–107.

————, "Elemente einer systemtheoretischen Betrachtung des Stoffwechsels," *Helgoländer wissenschaftliche Meeresuntersuchungen*, 14 (1966a), 4–24.

————, "Aktuelle Beiträge zur systemtheoretischen Behandlung des

Stoffwechsels. Netzwerk-, graphentheoretische und weitere Verfahren," *Studia biophysica*, 1 (1966b), 405–412.

———— and R. M. LOCKER, "Die Bedeutung experimenteller Variabler für die Abhängigkeit der Gewebsatmung von der Körpergrösse, III, Stimulation der Atmung und Auftrennung in Substratanteile," *Pflügers Arch. ges. Physiol.*, 274 (1962), 581–592.

LOEWE, S., "Die quantitativen Probleme der Pharmakologie, *Erg. Physiol.*, 27 (1928).

LORENZ, K., "Die angeborenen Formen möglicher Erfahrung," *Z. Tierpsychologie*, 5 (1943), 235.

LOTKA, A. J., *Elements of Physical Biology*, (1925), New York, Dover, 1956.

LUMER, H., "The Consequences of Sigmoid Growth Curves for Relative Growth Functions," *Growth*, I (1937).

LURIA, ALEKSANDR, *The Role of Speech in the Regulation of Normal and Abnormal Behavior*, New York, Pergamon Press, 1961.

LUTHE, WOLFGANG, "Neuro-humoral Factors and Personality," *Perspectives in Personality Theory*, Henry David and Helmut von Braken, editors, London, Tavistock, 1957.

MACCIA, E. STEINER, and G. S. MACCIA, *Development of Educational Theory Derived from Three Educational Theory Models*, Project 5-0638, Columbus, Ohio, The Ohio State Research Foundation, 1966.

MAGOUN, HORACE, *The Waking Brain*, Springfield, Illinois, Charles C. Thomas, 1958.

MALEK, E. et al., *Continuous Cultivation of Microorganisms*. Prague: Czech. Acad. Sci., 1958, 1964.

MANNING, HON. E. C., *Political Realignment—A Challenge to Thoughtful Canadians*, Toronto/Montreal, McClelland & Steward, Ltd., 1967.

MARTIN, A. W., and F. A. FUHRMAN, "The Relationship Between Summated Tissue Respiration and Metabolic Rate in the Mouse and Dog," *Physiol. Zool.*, 28 (1955), 18–34.

Mathematical Systems Theory, edited by D. Bushaw et al., New York: Springer, since 1967.

MATHER, K. F., "Objectives and Nature of Integrative Studies," *Main Currents in Modern Thought*, 8 (1951) 11.

MATSON, FLOYD, *The Broken Image*, New York, George Braziller, 1964.

MAY, ROLLO, ERNEST ANGEL and HENRI ELLENBERGER, editors, *Existence: A New Dimension in Psychiatry and Psychology*, New York, Basic Books, 1958.

MAYER, J., "Growth Characteristics of Rats Fed a Synthetic Diet," *Growth*, 12 (1948), 341–349.

McCLELLAND, C. A., "Systems and History in International Relations—Some Perspectives for Empirical Research and Theory," *General Systems*, 3 (1958), 221–247.

McNEILL, W., *The Rise of the West*, Toronto, The University of Toronto Press, 1963.

MEIXNER, J. R., and H. G. REIK, "Thermodynamik der irreversiblen Prozesse," *Handbuch der Physik*, Bd. III/2, S. Flügge, editor, Berlin, Springer Verlag, 1959, pp. 413–523.

MENNINGER, KARL, "The Psychological Aspects of the Organism Under Stress," *General Systems*, 2 (1957), 142–172.

————, HENRI ELLENBERGER, PAUL PRUYSER, and MARTIN MAYMAN, "The Unitary Concept of Mental Illness," *Bull. Menninger Clin.* 22 (1958), 4–12.

————, MARTIN MAYMAN and PAUL PRUYSER, *The Vital Balance*, New York, The Viking Press, 1963.

MERLOO, JOOST, *The Rape of the Mind*, Cleveland, The World Publishing Company, 1956.

MESAROVIC, M. D., "Foundations for a General Systems Theory," *Views on General Systems Theory*, M. S. Mesarović, editor, New York, John Wiley & Sons, 1964. 1–24.

METZGER, W., "Psychologie," *Wissenschaftliche Forschungsberichte, Naturwissenschaftliche Reihe*, 52 (1941).

MEUNIER, K., "Korrelation und Umkonstruktion in den Grössenbeziehungen zwischen Vogelflügel und Vogelkörper," *Biol. Gen.*, 19 (1951), 403–443.

MILLER, J. G. et al., "Symposium: Profits and Problems of Homeostatic Models in the Behavioral Sciences," *Chicago Behavioral Sciences Publications*, 1 (1953).

MILLER, JAMES, "Towards a General Theory for the Behavioral Sciences," *Amer. Psychol.*, 10 (1955), 513–531.

MILSUM, J. H., *Biological Control Systems Analysis*, New York, McGraw-Hill, 1966.

MINSKY, MARVIN L., *Computation, Finite and Infinite Machines*, Englewood Cliffs, N. J., Prentice-Hall, Inc., 1967.

MITTASCH, A., *Von der Chemie zur Philosophie—Ausgewählte Schriften und Vorträge*, Ulm, 1948.

MITTELSTAEDT, H., "Regelung in der Biologie," *Regelungstechnik*, 2 (1954), 177–181.

————, editor, *Regelungsvorgänge in der Biologie*, München, Oldenbourg, 1956.

MORCHIO, R., "Gli Organismi Biologici Come Sistemi Aperti Stazionari Nel Modello Teorico di L. von Bertalanffy," *Nuovo Cimento*, 10, Suppl., 12 (1959), 110–119.

MOSER, H. and O. MOSER-EGG, "Physikalisch-chemische Gleichgewichte im Organismus," *Einzeldarstellungen a.d. Gesamtgeb. d. Biochemie* (Leipzig), 4 (1934).

MUMFORD, L., *The Myth of the Machine*, New York, Harcourt, Brace, 1967.

MURRAY, HENRY, "The Personality and Career of Satan," *Journal of Social Issues*, 18 (1962), 36–54.

NAGEL, E., *The Structure of Science*, London, Routledge & Kegan Paul, 1961.

NAROLL, R. S., and L. VON BERTALANFFY, "The Principle of Allometry in Biology and the Social Sciences," *General Systems* 1 (1956), 76–89.

NEEDHAM, J., "Chemical Heterogony and the Groundplan of Animal Growth," *Biol. Rev.*, 9 (1934), 79.

NETTER, H., "Zur Energetik der stationären chemischen Zustände in der Zelle," *Die Naturwissenschaften*, 40 (1953), 260–267.

———, *Theoretische Biochemie*, Berlin, Springer, 1959.

NEUMANN, J. von, "The General and Logical Theory of Automata," *Cerebral Mechanisms in Behavior*, L. A. Jeffries, editor, New York, Wiley, 1951.

———, and O. MORGENSTERN, *Theory of Games and Economic Behavior*, Princeton, Princeton University Press, 1947.

NUTTIN, JOSEPH, "Personality Dynamics," in *Perspectives in Personality Theory*, Henry David and Helmut von Bracken, editors, London, Tavistock, 1957.

OPLER, MARVIN, *Culture, Psychiatry and Human Values*, Springfield, Ill., Charles C. Thomas, 1956.

OSTERHOUT, W. J. V., "The Kinetics of Penetration," *J. Gen. Physiol.*, 16 (1933).

———, "Bericht Über Vorträge auf dem 14. Internationalen Kongress für Physiologie, Rom 1932," *Die Naturwissenschaften* (1933).

———, and W. M. STANLEY, "The Accumulation of Electrolytes," *J. Gen. Physiol.*, 15 (1932).

PARETO, V., *Cours de l'economie politique*, Paris, 1897.

PATTEN, B. C., "An Introduction to the Cybernetics of the Ecosystem: The Trophic-Dynamic Aspect," *Ecology*, 40 (1959), 221–231.

PIAGET, JEAN, *The Construction of Reality in the Child*, New York, Basic Books, 1959.

PRIGOGINE, I., *Etude thermodynamique des phénomènes irréversibles*, Paris, Dunod, 1947.

———, "Steady States and Entropy Production," *Physica*, 31 (1965), 719–724.

PUMPIAN-MINDLIN, EUGENE, "Propositions Concerning Energetic-Economic Aspects of Libido Theory," *Ann. N.Y. Acad. Sci.*, 76 (1959), 1038–1052.

PÜTTER, A., "Studien zur Theorie der Reizvorgänge," I–VII, *Pflügers Archiv* 171, 175, 176, 180 (1918–20).

———, "Studien über physiologische Aehnlichkeit, VI. Wachstumsähnlichkeiten," *Pflügers Arch. ges. Physiol.*, 180 (1920), 298–340.

QUASTLER, H. (ed.), *Information Theory in Biology*, Urbana, The University of Illinois Press, 1955.

RACINE, G. E., "A Statistical Analysis of the Size-Dependence of Metabolism Under Basal and Non-Basal Conditions," Thesis, University of Ottawa, Canada, 1953.

RAPAPORT, D., *The Structure of Psychoanalytic Theory*, Psychol. Issues, Monograph 6, 2 (1960), 39–64.

RAPOPORT, A., "Outline of a Probabilistic Approach to Animal Sociology," I-III, *Bull. Math. Biophys.*, 11 (1949), 183–196, 273–281; 12 (1950), 7–17.

————, "The Promise and Pitfalls of Information Theory," *Behav. Sci.*, I (1956), 303–315.

————, "Lewis F. Richardson's Mathematical Theory of War," *General Systems*, 2 (1957), 55–91.

————, "Critiques of Game Theory," *Behav. Sci.*, 4 (1959a), 49–66.

————, "Uses and Limitations of Mathematical Models in Social Sciences," *Symposium on Sociological Theory*, L. Gross, editor, Evanston, Illinois, Row, Peterson, 1959b. pp. 348–372.

————, *Fights, Games and Debates*, Ann Arbor, University of Michigan Press, 1960.

————, "Mathematical Aspects of General Systems Theory," *General Systems*, 11 (1966), 3–11.

————, and W. J. HORVATH, "Thoughts on Organization Theory and a Review of Two Conferences," *General Systems*, 4 (1959), 87–93.

RASHEVSKY, N., *Mathematical Biophysics*, Chicago, The University of Chicago Press, 1938. 3rd. ed. 1960.

————, *Mathematical Biology of Social Behavior*, The University of Chicago Press, 1951.

————, "The Effect of Environmental Factors on the Rates of Cultural Development," *Bull. Math. Biophys.*, 14 (1952), 193–201.

————, "Topology and Life: In Search of General Mathematical Principles in Biology and Sociology," *General Systems*, I (1956), 123–138.

REIK, H. G., "Zur Theorie irreversibler Vorgänge," *Annalen d. Phys.*, 11 (1953), 270–284, 407–419, 420–428.; 13 (1953), 73–96.

RENSCH, B., *Neuere Probleme der Abstammungslehre*, 2nd edition, Stuttgart, Enke, 1954.

————, "Die Evolutionsgesetze der Organismen in naturphilosophischer Sicht," *Philosophia Naturalis*, 6 (1961), 288–326.

REPGE, R., "Grenzen einer informationstheoretischen Interpretation des Organismus," *Giessener Hochschulblätter*, 6 (1962).

RESCIGNO, A., "Synthesis for Multicompartmental Biological Models," *Biochem. Biophys. Acta.*, 37 (1960), 463–468.

————, and G. SEGRE, *Drug and Tracer Kinetics*, Waltham, Massachusetts, Blaisdell, 1966.

RIEGL, A., *Die Spätrömische Kunstindustrie, Nach den Funden in Österreich-Ungarn*. Wien, Hof- und Staatsdruekerei, 1901.

ROSEN, R., "A Relational Theory of Biological Systems," *General Systems*, 5 (1960), 29–44.

————, *Optimality Principles in Biology*, London, Butterworths, 1967.

ROSENBROCK, H. H., "On Linear System Theory," *Proceedings IEE*, 114 (1967), 1353–1359.

ROTHACKER, ERICH, *Die Schichten der Persönlichkeit*, 3rd edition, Leipzig, Barth, 1947.

ROTHSCHUH, K. E., *Theorie des Organismus*, 2nd edition, München, Urban/Schwarzenberg, 1963.

ROYCE, JOSEPH, R., *The Encapsulated Man*, New York, Van Nostrand, 1964.

RUESCH, J., "Epilogue," *Toward a Unified Theory of Human Behavior*, 2nd edition, R. R. Grinker, editor, New York, Basic Books, 1967.

RUSSELL, B., *Human Knowledge, Its Scope and Limits*, London, 1948.

SCHAFFNER, KENNETH F., "Antireductionism and Molecular Biology," *Science*, 157 (1967), 644–647.

SCHAXEL, J., *Grundzüge der Theorienbildung in der Biologie*, 2nd edition, Jena, Fischer, 1923.

SCHER, JORDAN, editor, *Theories of the Mind*, New York, The Free Press, 1962.

SCHILLER, CLAIRE, editor and translator, *Instinctive Behavior*, London, Methuen & Co., 1957.

SCHOENHEIMER, R., *The Dynamic State of Body Constituents*, 2nd edition, Cambridge, (Mass.) Harvard University Press, 1947.

SCHULZ, G. V., "Über den makromolekularen Stoffwechsel der Organismen," *Die Naturwissenschaften*, 37 (1950), 196–200, 223–229.

———, "Energetische und statistische Voraussetzungen für die Synthese der Makromoleküle im Organismus," *Z. Elektrochem. angew. physikal Chem.*, 55 (1951), 569–574.

SCOTT, W. G., "Organization Theory: An Overview and an Appraisal," in *Organizations: Structure and Behavior*, J. A. Litterer, editor, New York, John Wiley & Sons, 1963.

SELYE, H., *The Stress of Life*, New York, McGraw-Hill, 1956.

SHANNON, CLAUDE, and WARREN WEAVER, *The Mathematical Theory of Communication*, Urbana, University of Illinois Press, 1949.

SHAW, LEONARD, "System Theory," *Science*, 149 (1965), 1005.

SIMON, H. A., "The Architecture of Complexity," *General Systems*, 10 (1965), 63–76.

SKINNER, B. F., "The Flight From the Laboratory," *Theories in Contemporary Psychology*, Melvin Marx, editor, New York, The Macmillan Company, 1963.

SKRABAL, A., "Von den Simultanreaktionen," *Berichte der deutschen chemischen Gesellschaft* (A), 77 (1944), 1–12.

———, "Die Kettenreaktionen anders gesehen," *Monatshefte für Chemie*, 80 (1949), 21–57.

SKRAMLIK, E. von, "Die Grundlagen der haptischen Geometrie," *Die Naturwissenschaften*, 22 (1934), 601.

SMITH, VINCENT E. (ed.), *Philosophical Problems in Biology*, New York, St. John's University Press, 1966.

SOROKIN, P. A., *Contemporary Sociological Theories*, (1928), New York, Harper Torchbooks, 1964.

———. *Modern Historical and Social Philosophies* (1950), New York, Dover, 1963.

———. "Reply to My Critics," *Pitirim A. Sorokin in Review*, Philip Allen, editor, Durham, Duke University Press, 1963.

———, *Sociological Theories of Today*, New York/London, Harper & Row, 1966.

SPENGLER, O., *Der Untergang des Abendlandes*, vol. 1, Munich, Beck, 1922.

SPIEGELMAN, S., "Physiological Competition as a Regulatory Mechanism in Morphogenesis," *Quart. Rev. Biol.*, 20 (1945), 121.

SPRINSON, D. B., and D. RITTENBERG, "The Rate of Utilization of Ammonia for Protein Synthesis," *J. Biol. Chem.*, 180 (1949a), 707–714.

———, "The Rate of Interaction of the Amino Acids of the Diet with the Tissue Proteins," *J. Biol. Chem.*, 180 (1949b), 715–726.

STAGNER, ROSS, "Homeostasis as a Unifying Concept in Personality Theory," *Psychol. Rev.*, 58 (1951), 5–17.

STEIN-BELING, J. von, "Über das Zeitgedächtnis bei Tieren," *Biol. Rev.*, 10 (1935), 18.

STOWARD, P. J., "Thermodynamics of Biological Growth," *Nature, Lond.*, 194 (1962), 977–978.

SYZ, HANS, "Reflection on Group- or Phylo-Analysis," *Acta Psychotherapeutica*, 11 (1963), Suppl., 37–88.

SZENT-GYÖRGYI, A., "Teaching and the Expanding Knowledge," *Science*, 146 (1964), 1278–1279.

TANNER, JAMES, and BÄRBEL INHELDER, editors, *Discussions on Child Development*, vol. 4, London, Tavistock, 1960.

THOMPSON, J. W., "The Organismic Conception in Meteorology," *General Systems*, 6 (1961), 45–49.

THUMB, NORBERT, "Die Stellung der Psychologie zur Biologie: Gedanken zu L. von Bertalanffy's *Theoretischer Biologie*," *Zentralblatt für Psychotherapie*, 15 (1943), 139–149.

TOCH, HANS, and ALBERT HASTORF, "Homeostasis in Psychology: A Review and Critique," *Psychiatry: Journal for the Study of Inter-Personal Processes*, 18 (1955), 81–91.

TOYNBEE, A., *A Study of History*, vol. IX, London and New York, Oxford University Press, 1954.

———, *A Study of History*, vol. XII, *Reconsiderations*, London, New York: Oxford U. Press, 1961 (Galaxy), 1964.

TRIBIÑO, S. E. M. G. de, "Una Nueva Orientación de la Filosofía Biológica: El Organicismo de Luis Bertalanffy, Primer premio 'Miguel Cané,' ", *Cursos y Conferencias* (Buenos Aires), 28 (1946).

TRINCHER, K. S., *Biology and Information: Elements of Biological Thermodynamics*, New York, Consultants Bureau, 1965.

TSCHERMAK, A. von, *Allgemeine Physiologie*, 2 vols. Berlin, Springer, 1916, 1924.

TURING, A. M., "On Computable Numbers, with an Application to the Entscheidungsproblem," *Proc. London. Math. Soc.*, Ser. 2, 42 (1936).

UEXKÜLL, J. von, *Umwelt und Imnenwelt der Tiere*, 2nd edition, Berlin, Springer, 1920.

————, *Theoretische Biologie*, 2nd edition, Berlin, Springer, 1929.

————, and G. KRISZAT, *Streifzüge durch die Umwelten von Tieren und Menschen*, Berlin, Springer, 1934.

UNGERER, E., *Die Wissenschaft vom Leben. Eine Geschichte der Biologie, Bd. III*, Freiburg/München, Alber, 1966.

VICKERS, G., "Control, Stability, and Choice," *General Systems*, II (1957), 1–8.

VOLTERRA, V., *Leçons sur la Théorie Mathématique de la Lutte pour la Vie*, Paris, Villars, 1931.

WAGNER, RICHARD, *Probleme und Beispiele Biologischer Regelung*, Stuttgart, Thieme, 1954.

WAHL, O., "Neue Untersuchungen über das Zeitgedächtnis der Tiere," *Z. Vergl. Physiol.*, 16 (1932), 529.

WATT, K. E. F., "The Choice and Solution of Mathematical Models for Predicting and Maximizing the Yield of a Fishery," *General Systems*, 3 (1958), 101–121.

WEAVER, W., "Science and Complexity," *American Scientist*, 36 (1948), 536–544.

WEISS, P., "Experience and Experiment in Biology," *Science*, 136 (1962a), 468–471.

————, "From Cell to Molecule," *The Molecular Control of Cellular Activity*, J. M. Allen, ed., New York, 1962b.

WERNER, G., "Beitrag zur mathematischen Behandlung pharmakologischer Fragen," *S.B. Akad. Wiss. Wien., Math. Nat. Kl.* 156 (1947), 457–467.

WERNER, HEINZ, *Comparative Psychology of Mental Development*, New York, International Universities Press, 1957a.

————, "The Concept of Development from a Comparative and Organismic Point of View," *The Concept of Development*, Dale Harris, editor, Minneapolis, University of Minnesota Press, 1957b.

WHATMOUGH, J., "Review of Logic and Language (Second Series)," A. G. N. Flew, editor, *Classical Philology*, 50 (1955), 67.

WHITEHEAD, A. N., *Science and the Modern World*, Lowell Lectures (1925), New York, The Macmillan Company, 1953.

WHITTACKER, R. H., "A Consideration of Climax Theory: The Climax as a Population and Pattern," *Ecol. Monographs*, 23 (1953), 41–78.

WHORF, B. L., *Collected Papers on Metalinguistics*, Washington, Foreign Service Institute, Department of State, 1952.

————, *Language, Thought and Reality: Selected Writings of B. L. Whorf*, John Carroll, editor, New York, John Wiley & Sons, 1956.

WHYTE, LANCELOT, *The Unconscious before Freud*, New York, Basic Books, 1960.

WIENER, NORBERT, *Cybernetics*, New York, John Wiley & Sons, 1948.

WOLFE, HARRY B., "Systems Analysis and Urban Planning—The San Francisco Housing Simulation Model," *Trans. N.Y. Acad. Sci.*, Series II, 29:8 (June 1967), 1043–1049.

WOODGER, J. H., "The 'Concept of Organism' and the Relation between Embryology and Genetics," *Quart. Rev. Biol.*, 5/6 (1930–31), 1–3.

————, *The Axiomatic Method in Biology*, Cambridge, 1937.

WORRINGER, W., *Abstraktion und Einfühlung*, Munich, Piper, 1908.

————, *Formprobleme der Gotik*, Munich, Piper, 1911.

YOURGRAU, W., "General System Theory and the Vitalism-Mechanism Controversy," *Scientia* (Italy), 87 (1952), 307.

ZACHARIAS, J. R., "Structure of Physical Science," *Science*, 125 (1957), 427–428.

ZEIGER, K., "Zur Geschichte der Zellforschung und ihrer Begriffe," *Handb. d. Allgem. Pathol.*, F. Büchner, E. Letterer, and F. Roulet, editors, Bd. 2, T. 1, 1–16, 1955.

ZERBST, E., "Eine Methode zur Analyse und quantitativen Auswertung biologischer steady-state Übergänge," *Experientia*, 19 (1963a), 166.

————, "Untersuchungen zur Veränderung energetischer Fliessgleichgewichte bei physiologischen Anpassungsvorgängen," I, II, *Pflügers Arch. ges. Physiol.*, 227 (1963b), 434–445, 446–457.

————, *Zur Auswertung biologischer Anpassungsvorgänge mit Hilfe der Fliessgleichgewichtstheorie. Habilitationsschrift.* Berlin, Freie Universität, 1966.

————, *Eine Analyse der Sinneszellfunktion mit Hilfe der von Bertalanffy-Fliessgleichgewichtstheorie*, Berlin, Freie Universität, (in press).

————, C. HENNERSDORF, and H. VON BRAMANN, "Die Temperaturadaptation der Herzfrequenz und ihre Analyse mit Hilfe der Fliessgleichgewichttheorie," 2nd International Biophysics Congress of the International Organization for Pure and Applied Biophysics, Vienna, Sept. 5–9, 1966.

ZUCKER., L., L. HALL, M. YOUNG, and T. F. ZUCKER, "Animal Growth and Nutrition, With Special Reference to the Rat," *Growth*, 5 (1941a), 399–413.

————, "Quantitive Formulation of Rat Growth," *Growth*, 5 (1941b), 415–436.

ZUCKER, T. F., L. HALL, M. YOUNG, and L. ZUCKER, "The Growth Curve of the Albino Rat in Relation to Diet," *J. Nutr.*, 22 (1941), 123–138.

ZUCKER, L. and T. F. ZUCKER, "A Simple Weight Relation Observed in Well-Nourished Rats," *J. gen. Physiol.* 25 (1942), 445–463.

ZWAARDEMAKER, H., "Die im ruhenden Körper vorgehenden Energiewanderungen," *Erg. Physiol.*, 5 (1906).

————, "Allgemeine Energetik des tierischen Lebens (Bioenergetik)," *Handbuch der normalen und pathologischen Physiologie*, I (1927).

SUGGESTIONS FOR FURTHER READING

The following list is intended for further study in general system theory as defined in this book, and its major fields of application. For this reason, only a few representative examples are cited from the large literature in fields such as cybernetics, information, game and decision theories, irreversible thermodynamics, systems analysis and engineering, etc.

General, Mathematics of General System Theory

"Biologische Modelle," Symposium, *Nova Acta Leopoldina* (Halle, Germany), 1969. (Articles L. von Bertalanffy, H. Drischel, Benno Hess, etc.).

BOGUSLAW, W., *The New Utopians*, Englewood Cliffs (N.J.), Prentice-Hall, 1965.

BUCKLEY, W. (ed.), *Modern Systems Research for the Behavioral Scientist. A Sourcebook*, Chicago, Aldine Publishing Co., 1968.

General Systems, L. von Bertalanffy and A Rapoport (eds.), Society for General Systems Research, 2100 Pennsylvania Avenue, N.W., Washington, D.C., 15 vols. since 1956.

GORDON, Jr., Charles K., *Introduction to Mathematical Structures*, Belmont (Cal.), Dickenson, 1967.

JONES, R. D. (ed.), *Unity and Diversity*, Essays in Honor of Ludwig von Bertalanffy, New York, Braziller, 1969. (Articles A. Auersperg, W. Beier and R. Laue, R. Brunner, A. Koestler, A. Rapoport, R. B. Zuñiga, etc.).

KLIR, G. J., *An Approach to General Systems Theory*, Princeton (N.J.), Nostrand, 1968.

MACCIA, Elizabeth Steiner, and George S. MACCIA, *Development of Educational Theory Derived from Three Educational Theory Models*, Columbus (Ohio), The Ohio State University, 1966.

MESAROVIĆ, M. D., *Systems Research and Design; View on General Systems Theory*, New York, Wiley, 1961 and 1964; *Systems Theory and Biology*, New York, Springer-Verlag, 1968.

System Theory, Proceedings of the Symposium on, Brooklyn (N.Y.), Polytechnic Institute, 1965.

Texty ke studiu teorie řízení. Řada: Teorie systému a jeji aplikace, Prague, Vysoka Škola Politická ÜV ḰSČ, 1966.

282 GENERAL SYSTEM THEORY

Biophysics

BEIER, Walter, *Einführung in die theoretische Biophysik*, Stuttgart, G. Fischer, 1965.

BERTALANFFY, L. von, *Biophysik des Fliessgleichgewichts*, translated by W. H. Westphal, Braunschweig, Vieweg, 1953. Revised ed. in preparation.

BRAY, H. G. and K. WHITE, "Organisms as Physico-Chemical Machines," *New Biology*, 16 (1954) 70–85.

DOST, F. H., *Grundlagen der Pharmakokinetik*, 2. Aufl., Stuttgart, Thieme, 1968.

FRANKS, Roger G. E., *Mathematical Modeling in Chemical Engineering*, New York, Wiley, 1967.

Quantitative Biology of Metabolism, International Symposia, A. Locker and O. Kinne (eds.), Helgoländer Wissenschaftliche Meeresuntersuchungen, 9, 14 (1964), (1966).

RESCIGNO, Aldo and Giorgio SEGRE, *Drug and Tracer Kinetics*, Waltham (Mass.), Blaisdell, 1966.

YOURGRAU, Wolfgang, A. VAN DER MERWE and G. RAW, *Treatise on Irreversible and Statistical Thermophysics*, New York, Macmillan, 1966.

Biocybernetics

BAYLISS, L. E., *Living Control Systems*, San Francisco, Freeman, 1966.

DiSTEFANO, III, Joseph J., A. R. STUBBERUD, and I. J. WILLIAMS, *Schaum's Outline of Theory and Problems of Feedback and Control Systems*, New York, Schaum, 1967.

FRANK, L. K. et al., *Teleological Mechanisms*, N. Y. Acad. Sc., 50 (1948).

GRODINS, Fred Sherman, *Control Theory and Biological Systems*, New York, Columbia University Press, 1963.

HASSENSTEIN, Bernhard, "Die bisherige Rolle der Kybernetik in der biologischen Forschung," *Naturwissenschaftliche Rundschau*, 13 (1960) 349–355, 373–382, 419–424.

———, "Kybernetik und biologische Forschung," *Handbuch der Biologie*, L. von Bertalanffy and F. Gessner (eds.), Bd. I, Frankfurt a.M., Athenaion, 1966, pp. 629–730.

KALMUS, H. (ed.), *Regulation and Control in Living Systems*, New York, Wiley, 1966.

MILSUM, John H., *Biological Control Systems Analysis*, New York, McGraw-Hill, 1966.

WIENER, N., *Cybernetics*, New York, Wiley, 1948.

Ecology and Related Fields

BEVERTON, R. J. H., and S. J. HOLT, *On the Dynamics of Exploited Fish Populations, Fishery Investigation*, Ser. II, vol. XIX. London, Her Majesty's Stationery Office, 1957.

WATT, Kenneth E. F., (ed.), *Systems Analysis in Ecology*, New York, Academic Press, 1966.

Psychology and Psychiatry

BERTALANFFY, L. von, *Robots, Men and Minds*, New York, Braziller, 1967.

GRAY, W., N. D. RIZZO and F. D. DUHL (eds.), *General Systems Theory and Psychiatry*, Boston, Little, Brown, 1968.

GRINKER, Roy R. (ed.), *Toward a Unified Theory of Human Behavior*, 2nd ed., New York, Basic Books, 1967.

KOESTLER, A., *The Ghost in the Machine*, New York, Macmillan, 1968.

MENNINGER, K., with M. MAYMAN, and P. PRUYSER, *The Vital Balance*, New York, Viking Press, 1963.

Social Sciences

BUCKLEY, W., *Sociology and Modern Systems Theory*, Englewood Cliffs (N.J.), Prentice-Hall, 1967.

DEMERATH III, N. J., and R. A. PETERSON (eds.), *System, Change, and Conflict. A Reader on Contemporary Sociological Theory and the Debate over Functionalism*, New York, Free Press, 1967.

HALL, Arthur D., *A Methodology for Systems Engineering*, Princeton (N.J.), Nostrand, 1962.

PARSONS, Talcott, *The Social System*, New York, Free Press, 1957.

SIMON, Herbert A., *Models of Man*, New York, Wiley, 1957.

SOROKIN, P. A., *Sociological Theories of Today*, New York, London, Harper & Row, 1966.

Addenda (1971)

BEIER, W., *Biophysik*. 3rd edition. Leipzig: Georg Thieme, 1968.

BEIER, W. and W. LAUE. "On the Mathematical Formulation of Open Systems and Their Steady States." In *Unity Through Diversity*, loc cit, Book II.

BERTALANFFY, L. von, "Chance or Law." In *Beyond Reductionism*, loc cit.

BERTALANFFY, L. von, "The History and Status of General System Theory." In *Trends in General Systems Theory*, loc cit.

Beyond Reductionism. Edited by A. Koestler and J. R. Smythies. London, New York: Hutchinson, 1969.

HAHN, W., *Theory and Application of Liapunov's Direct Method*. Englewood Cliffs, N.J.: Prentice-Hall, 1963.

HARVEY, D., *Explanation in Geography*. London: Arnold, 1969.

Hierarchical Structures, Edited by L. L. Whyte, A. G. Wilson and D. Wilson, New York: Elsevier, 1969.

Journal of General Systems. Edited by G. J. Klir and others. Starting 1972.

284

KOESTLER, A., "The Tree and the Candle." In *Unity Through Diversity*, loc cit, Book II.

LA SALLE, J. and S. LEFSCHETZ, *Stability by Liapunov's Direct Method*. New York, London: Academic Press, 1961.

LASZLO, E., "Systems and Structures—Toward Bio-Social Anthropology." In *Unity Through Diversity*, loc cit, Book IV.

LASZLO, E., *Introduction to Systems Philosophy*. London, New York: Gordon and Breach, in press.

MILLER, J. G., "Living Systems: Basic Concepts." In *General Systems Theory and Psychiatry*, loc cit.

MILSTEIN, M. and J. BELASCO, *Educational Administration of the Behavioral Sciences: A Systems Perspective*. Boston: Allyn and Bacon, in press.

ROSEN, R., "Two-Factor Models, Neural Nets in Biochemical Automata." *Journal of Theoretical Biology*, 15, (1967), 282–297.

ROSEN, R., *Dynamical System Theory in Biology*, Vol. I: *Stability Theory and its Applications*. New York: Wiley, 1970.

ROSEN, R., "A Survey of Dynamical Descriptions of Systems Activity." In *Unity Through Diversity*, loc cit, Book II.

SCHWARZ, H., *Einführung in die moderne Systemtheorie*. Braunschweig: Vieweg, 1969.

Systems Thinking. Edited by F. E. Emery. London: Penguin Books, 1969.

Trends in General Systems Theory. Edited by G. J. Klir. New York: Wiley, 1971.

Unity Through Diversity. A Festschrift in Honor of Ludwig von Bertalanffy. Edited by W. Gray and N. Rizzo. Espec. Book II: "General and Open Systems," Book IV: "General Systems in the Behavioral Sciences." London, New York: Gordon and Breach, 1971.

WEISS, P. A., *Life, Order and Understanding*. The Graduate Journal, The University of Texas, vol. VIII, Supplement, 1970.

WILBERT, H., "Feind-Beute-Systeme in kybernetischer Sicht," in *Oecologia* (Berlin) 5, (1970), 347–373.

Index

Ackoff, R. L., 9, 91, 100, 101
Actions of animal and human body, feedback mechanisms in regulation of, 43–44
Active personality system, model of man as, 192–93
Actuality principle, 116
Adams, H., 159
Adaptiveness, model for, 46
Adolph, E. F., 171
Afanasjew, W. G., 12
Alexander, Franz, 207
Allesch, G. J. von, 229
Allometric equation: definition, 63–65; in biology, 163–71 (tables, figs.); in social phenomena, 103
Allport, Floyd, 205
Allport, Gordon W., 193, 205, 206, 207, 208, 209, 212, 216
American Association for the Advancement of Science, 15
American Psychiatric Association, 7
Analogies in science: definition, 84, 85; value 35–36
Analytical procedure in science, 18–19
Anderson, Harold, 205
Anschütz, G., 231
Appleby, Lawrence, 216
Archaeology, process-school of, 9
Arieti, Silvano, 194, 205, 207, 211, 212, 214–15, 216, 217

Aristotle, 70, 79, 212, 216, 223, 225, 231, 246
Arrow, K. J., 113, 115
Ashby, W. R., 25, 46, 94, 96–99, 244
Atomic energy development, 118, 187
Attneave, F., 100
Ausubel, David P., 194
Automata, theory of, 22, 25, 141
Automation Revolution, 187

Backman, G., 231
Bavink, B., 76, 243
Bayliss, L. E., 22
Beadle, G. W., 152
Beckner, M., 12
Beer, S., 96
Behavior: adaptiveness, purposiveness, and goal-seeking in, 45–46, 79, 92, 131; unitary and elementalistic conceptions of, 70–71; stimulus-response (S-R) scheme, 107, 188–89, 191, 193, 209; and principle of rationality, 115–16; and environmentalism, 189–90; equilibrium principle in, 190; principle of economy in, 190; see also Human behavior
Behavioristic psychology, 7, 107, 187
Beier, W., 145, 149
Bell, E., 100
Bendmann, A., 12
Benedict, Ruth, 201, 219

Bentley, A. F., 41
Berg, K., 181
Berlin, Sir Isaiah, 9, 113, 114
Berlyne, D. E., 209, 212
Bernal, J. D., 5, 12
Bernard, Claude, 12
Bernstein, A., 205
Bertalanffy, Felix D., 147
Bertalanffy, Ludwig von, 6, 7, 9, 11,
    12, 13, 14, 38, 46, 67, 70, 72, 73,
    76, 77, 79, 89, 94, 95, 98, 99, 102,
    103, 104, 106, 116, 118, 121, 135,
    136, 141, 142, 145, 148, 151, 153,
    158, 159, 171, 174, 181, 183, 207,
    208, 209, 210, 212, 213, 215, 216,
    217, 218, 219, 220, 221, 226, 229,
    231, 236, 242, 244, 247, 248, 249,
    257, 259
Bethe, Albert, 208
Beverton, R. J. H., 104, 148
Biculturalism in Canada, 202
Biocoenoses, 68, 138, 149
Biological equilibria, theory of, 32, 47
Biological relativity of categories,
    227–32
Biologism, 88, 118
Biology: molecular, 6; higher levels of
    organization, 6, 28, 31; organismic
    conception in, 6, 12, 31, 89, 102–3,
    205, 208,259; mechanism-vitalism
    controversy in, 89; aspects of
    system theory in, 155–85
Blandino, G., 12
Blasius, W., 158
Bleuler, Eugen, 208, 214–15, 218
Blood, as open system, 148
Bode, H., 49–50
Boffey, Philip M., 4
Boguslaw, W., 3, 10
Bohr, Niels Henrik David, 182
Boltzmann, Ludwig, 30, 151, 152
Borelli, Giovanni Alfonso, 140
Boulding, K. E., 14, 27, 47, 103, 104,
    199
Bradley, D. F., 96, 147
Brainwashing, 191
Brave New World (A. Huxley), 10,
    52, 108, 118
Bray, H. G., 102
Bray, J. R., 102
British Ministry of Agriculture and
    Fisheries, 104
Brody, S., 165, 231
Bronowski, J., 23
Bruner, Jerome, 212

Brunner, R., 149
Brunswik, Egon, 205
Buckley, W., 8, 17, 196
Bühler, Charlotte, 107, 205, 207
Bühler, K., 209
Burton, A. C., 141, 144, 147
Butenandt, A., 158

Calvin, M., 96, 147
Cannon, W. B., 12, 16, 23, 78, 161, 211
Cantril, Hadley, 194, 212
Carlyle, Thomas, 111
Carmichael, Leonard, 209
Carnap, R., 86, 87
Carter, L. J., 4
Cartesian dualism between matter
    and mind, 220
Casey, E. J., 161
Cassirer, Ernst, 194, 212, 216
Categories: Kant's table of, 45; theory
    of (N. Hartmann), 85–86;
    introduction of new, in scientific
    thought and research, 10, 18, 92,
    94; relativity of, and Whorfian
    hypothesis, 222–27; biological
    relativity of, 227–32; cultural
    relativity of, 232–38; relativity
    of, and the perspectivistic view,
    239–48
Causal laws, 44–45
Center for Advanced Study in the
    Behavioral Sciences (Palo Alto),
    14
Centralization: definition, 71–74; in
    psychopathology, 213–14
Chance, B., 147, 163
Chemical equilibria, 121–125
Chemodynamic machine, 140
Chomsky, N., 189
Chorley, R. J., 102
"Classical" system theory, 19–20
Clausius, Rudolf J. E., 151
Closed and open systems, 39–41,
    121–25, 141
Coghill, G. E., 208
Commoner, B., 12
Communication, theory of, 41; and
    flow of information, 41–42; and
    concept of feedback, 42–44 (fig.)
Communication engineering, 22
Compartment theory, 21, 144
Competition between parts: equations
    defining, 63–66, 149; "wholes"
    based upon, 66, 91; and
    allometry, 163–164

Complexes of "elements," 54–55 (*fig.*)
Computer technology and cybernetics, 15
Computerization and simulation, as approach in systems research, 20–21 (*table*), 144
Conceptual experimentation at random, 101
Conditioning, 52, 189
Conflict and Defence (Boulding), 199
Conklin, E. W., 211
Constitutive and summative characteristics, 54–55
Control engineering and power engineering, 3
Convergence of research, 243
Copernican Revolution, 99
Cowdry, Edmund, 211
Crime and Criminologists (Anon.), 207
Critique of Practical Reason (Kant), 186
Cultural relativity of categories, 232–38
Culture: laws in development of, 199; concept of, 201–2; as psycho-hygienic factor, 218; multiplicity of, 235
Cummings, J., 216
Cusa, Nicholas of, 11, 248
Cybernetic machines, 140
Cybernetics: development of, in technology and science, 16–17, 23, 101; as part of general theory of systems, 17, 21–22; feedback mechanisms in, 44, 78, 90, 150, 161; and open systems, 149–50
Cybernetics (Wiener), 15

Damude, E., 7
D'Ancona, V., 56, 57, 76, 80, 133, 134, 138
Darwinism, 24, 152, 154
De-anthropomorphization in science, 242–44, 247, 248
Decision theory approach to systems problems, 22, 90, 100, 114, 115, 198–199
Decline of the West, The (Spengler), 118, 203
Decline of the West, as accomplished fact, 204
De ludo globi (Nicholas of Cusa), 11
Demerath, N. J., 196

Denbigh, K. G., 144, 151
Descartes, René, 19, 140, 212, 235, 240
De-Shalit, A., 5
Determinism, 114, 221
Differentiation, principle of, in psychopathology, 211–13; *see also* Segregation Diffusion, 126–27; cultural, 201
Directed graph (digraph) theory, 21
Directiveness of processes, 16–17, 45–46, 78, 92
Dobzhansky, T., 12
Donnan, F. G., 57, 134
Dost, F. H., 148, 158, 175
Driesch, Hans, 26, 27, 40, 72, 133, 144
Drischel, H., 19
Druckery, H., 148
Drug kinetics, 56, 138, 148
Dubos, R., 12
Dunn, M. S., 180
Dynamic ecology, 102
"Dynamic State of Body Constituents" (Schönheimer), 160
Dynamic equilibrium, 131; *see also* Steady state
Dynamic teleology, 78–79
Dynamics of biological populations, theory of, 32

Ecology, theory of, 32, 47, 102
Economics and econometrics, 32
Economy, principle of, in human behavior, 190
Eddington, Sir Arthur Stanley, 151
Education, general system theory in, 49–51, 193
"Education of Scientific Generalists, The" (Bode, *et al.*), 49–50
Ego boundary, in psychopathology, 215
Einstein, Albert, 154, 225, 247
Elsasser, W. M., 25, 161
Emergence, 55
Entropy, 39, 41, 143, 144, 145, 151, 152, 159; *see also* Thermodynamics
Environmentalism, principle of, 189–90, 191
Equifinality, 46, 102, 136, 144; definition, 40, 132–134; of growth, 142 (*fig.*), 148–49
Equilibrium principle in human behavior, 190
Euler, Leonhard, 75

Evolution: and contrast between
    wholeness and sum in, 70;
    synthetic theory of, 152, 153, 187
Excitation, phenomena of, and open
    systems concept, 137, 138
Existentialism, 109, 193
Explanation in principle, 36, 47, 106,
    113
Exponential law, 61–62 (fig.) , 82
Eysenck, Hans, 214

Factor analysis, 90
Fagen, R. E., 95
Fearing, F., 222
Fechner, Gustav Theodor, 107
Feedback: concept of, 42–44 (fig.) , 46,
    150; and homeostasis, 43, 78, 101,
    150, 160–63, 184; and cybernetics,
    44, 78, 79, 150, 161; criteria of
    control systems, 161–63
Fights, Games and Debates
    (Rapoport) , 199
Finality systems, 75–77, 91, 131; types
    of, 77–80
Fitness, and teleology, 77, 79
Flannery, Kent V., 9
Foerster, H. von, 163
Food and Agricultural Organization
    of U.N., 104
Foster, C. A., 151
Foundation for Integrated Education,
    50
"Four R's of Remembering"
    (Pribram), 211n.
Frank, L. K., 16, 17, 78
Frankl, Victor, 211, 216, 217, 219
Franks, R. G. E., 144
Free will, 114, 115, 116, 221
Freeman, Graydon, 210
Freud, Sigmund, 105, 107, 115, 189,
    190, 194, 212, 214, 216
Friedell, Egon, 192
Fuhrmann, F. A., 168
Functionalism, and sociological
    theory, 196
Future, system-theoretical view of,
    203–4

Galileo, 19, 182, 186
Gallup polls, 116, 198
Game theory, 15, 22, 23, 90, 100, 110,
    114, 115, 198, 238
Garavaglia, C., 147
Gause, G. F., 47, 56, 103

Gauss, Karl Friedrich, 90
Gazis, Denos C., 20
Geertz, Clifford, 212
General system theory: history of,
    10–17, 89–90; trends in, 17–29;
    approaches to methodological
    problems of, 19–23;
    axiomatization, 21; quest for,
    30–36; trends toward generalized
    theories in multiple fields, 32;
    postulation of new discipline of,
    32, 37, 90; meaning of, 32–33;
    and isomorphisms in different
    fields, 33–34; as a general science
    of organization and wholeness,
    34, 36–37; objections to, 35–36;
    aim defined, 15, 38; examples,
    38–49; and unity of science,
    48–49, 86–88, 253; integrative
    function of, 48–49; in education,
    49–51, 193; motives leading to
    postulate of, 91–94; advances in,
    99–119; see also System, etc.
General systems research, methods in,
    94–99; empirico-intuitive method,
    95–96; deductive approach, 96–99
General Systems, Yearbooks of Society
    for General Systems Research, 15
Geomorphology, 102
Gerard, Ralph W., 15, 34
Gessner, F., 103
Gestalt psychology, 6, 31, 208
Geyl, Peter, 110
Ghost in the Machine, The
    (Koestler) , 214n.
Gibson, J. J., 211n.
Gilbert, Albin, 213
Gilbert, E. N., 22
Glansdorff, P., 151
Glasperlenspiel (Hesse), 11
Global nature of our civilization, 204
Goal-seeking behavior, 16, 43–44, 45,
    79, 92, 131, 150
Goethe, Johann Wolfgang von, 145,
    249
Goldstein, Kurt, 105, 207, 208, 216,
    217
Graph theory approach to systems
    problems, 90, 211
Gray, William, 7
Grinker, Roy R., 7
Grodin, F. S., 161
Gross, J., 77
Group theory, 237–38

Growth: general systems equations, 60–63 *(fig.)*; exponential, 61–62; logistic, 62–63; gr. equations (model) after Bertalanffy, 103, 135–36, 148, 171–84; relative, 103, 149, *see also* Allometric equation; equifinality of, 142 *(fig.)* , 148–49

Guerra, E., 171

Günther, B., 171

Hacker, Frederick, 207

Hahn, Erich, 6, 10

Haire, M., 96, 103, 114, 118

Hall, A. D., 91, 95, 105

Hall, C. S., 105

Hart, H., 26

Hartmann, E. von, 77

Hartmann, M., 124

Hartmann, Nicolai, 72, 86

Harvey, William, 140

Hastorf, Albert, 211

Hayek, F. A., 36, 113

Hearn, G., 95

Hearon, J. F., 144

Hebb, Donald O., 106, 209

Hecht, S., 137, 148

Hegel, Georg Wilhelm Friedrich, 11, 110, 198, 199

Heisenberg, Werner, 31

Hemmingsen, A. M., 183

Hempel, C. G., 12

Henry, Jules, 206

Heraclitus, 160, 246, 247

Hering, Ewald, 137

Herrick, Charles, 209, 216

Hersh, A. H., 62, 103

Herzberg, A., 13

Hess, B., 20, 144, 147, 163

Hess, W. R., 16

Hesse, Hermann, 11

Heterostasis, 23

Hierarchic order in general systems theory, 27–29 *(table)* , 74, 213

Hill, A. V., 137

Hippocrates, 237

Historical inevitability, 8–9, 113–114, 118

History: cyclic model of, 118, 203, 204; impact of systems thinking on conception of, 8–9; theoretical, 109–19, 197–203; nature of historical process, 200–201; organismic theory of, 202–203

Hoagland, H., 230

Höber, R., 135

Höfler, Otto, 81*n.*

Hoijer, H., 226

Holst, Erich von, 16, 106, 209

Holst, S. J., 104, 148

Homeostasis: Cannon's concept of, 12, 16, 23, 78, 161; and feedback, 43, 78, 101, 150, 160–63, 184; in psychology and psychopathology, 210–11

Homology, logical, 84–85

Hook, Sidney, 220

Horvath, W. J., 117

Human behavior: theory of, 105; robot model of, 188, 190–91, 206; aspects of, outside of physical laws, 199; *see also* Behavior

Human element, as component in systems engineering, 10

Human engineering, 91

Human society: application of general system theory to, 47–48; laws and science of, 51–52; evaluation of man as an individual in, 52–53; and statistical laws, 116, 198

Humanistic psychology, 193

Humboldt, Wilhelm von, 194, 212, 232

Huxley, Aldous, 49, 52, 231

Huxley, Sir Julian, 149

Hypothetic-deductive approach, 198

Ibn-Khaldun, 11

Idiographic method in history, 8, 110, 111, 114, 198

Inanimate and animate nature, apparent contrast between 40–41, 139

"Immense" numbers, problem of, in systems theory, 25–27

Individualization within system, 71–74

Industrial chemistry, 122, 142

Industrial Revolution, 118, 186–87

Information theory, 15, 22, 90, 93, 94, 100, 151, 152, 163, 198, 245

Inhelder, Bärbel, 221

Instinct, theory of, 106

Institute for Advanced Study in Princeton, 5

Integrative function of general system theory, 48–49

"Integrative Studies for General Education" (Mather), 50
Interdisciplinary theory: implications of, 48–49; basic principles of, 51; and new conceptual models, 93–94
Isomorphisms: in different fields, 33–34, 48–49, 88, 103; in science, 80–86

Jeffries, L. A., 27
Jones, R. W., 161
Jung, Carl, 105
Jung, F., 102

Kafka, Franz, 77
Kalmus, H., 22, 230
Kamarýt, J., 12
Kanaev, I. I., 12
Kant, Immanuel, 45, 101, 186–87, 225, 227, 229, 232, 240
Keiter, F., 95, 103
Kelvin, William Thomson, 40
Kinetics, 13, 56, 120, 141, 150, 156, 159
Kleiber, M., 165
Kluckhohn, C., 201, 224, 249
Kment, H., 101, 161
Koestler, Arthur, 29, 212, 214n.
Köhler, W., 11, 131, 208
Kottje, F., 124
Krech, David, 41, 105, 205
Kremyanskiy, V. I., 96
Kriszat, G., 228
Kroeber, A. L., 156, 198, 201
Kubie, Lawrence, 217
Kuepfmüller, K., 148
Kuhn, T. S., 18, 24, 201

La Barre, W., 224, 225
Landois-Rosemann textbook, 158
Langer, S., 216
Lapicque, L., 137
Laplace, Pierre Simon, 21, 26, 30, 87, 113
Lashley, K., 26, 208
Lau, C., 147
Laue, R., 145
Laws of nature, modern concept of, 113
Le Chatelier's principle in physical chemistry, 75, 80, 131
Lecomte du Noüy, P., 231
Lehmann, G., 165

Leibniz, Gottfried Wilhelm, 11, 250
Leighton, D., 224
Lennard H., 205
Lenz's rule of electricity, 75, 80
Lersch, P., 213
Lewada, J., 10
Lindzey, G., 105
Linguistic determination of categories of cognition, Whorfian hypothesis of, 194, 222–27
Linguistic systems, diversity of, and reevaluation of scientific concepts, 225
Living organism: as open system, 32, 39, 44, 121–23, 141, 156–60 (fig.), 191; and dynamic interplay of processes, 44; and machine conception, 139–41; biophysics of, 142, 158
Llavero, F., 207
Locker, A., 25, 144, 145, 166, 168, 169
Loewe, S., 138, 148
Logistic curve, 62–63 (fig.)
Lorenz, K., 106, 240, 249
Lotka, A. J., 11, 32, 47, 56, 103
Lumer, H., 64
Luria, Aleksandr, 216
Luthe, Wolfgang, 213

Maccia, E. S. and G. S., 21
"Macrohistory," 199
Magoun, Horace, 209
Malek, E., 149
Malthusian law of population, 48, 62, 104
Man: image of, in contemporary thought, 6, 188–92, 194; role of, in the Big System, 10; as the individual, ultimate precept of theory of organization, 52–53; system concept in sciences of, 186–204; model of, as robot, 188, 190–91, 194, 205, 206; pecuniary conception of man, 206; as active personality system, 207; special position of, in nature, 257–258
Man-machine systems, 91
Manipulative psychology, 206–207
Manning, Hon. E. C., 4
Martin, A. W., 168
Marx, Karl, 11, 110, 198, 199
Maslow, A. H., 105, 109, 193, 207
Mass action, law of, 120, 122, 125
Mass behavior, 114

Mass civilization, 204

Mass suggestion, methods of, 52

Materialism, 94

Mathematical approaches in general systems theory, 19–23, 38, 90–91

Mathematical models, advantages of, 24

Mathematical Systems Theory journal, 15

Mather, K. F., 50

Matson, Floyd, 206, 212

Maupertuis, Pierre Louis Moreau de, 75

May, Rollo, 218

Mayer, J., 180

McClelland, C. A., 118

McCulloch, W. S., 25

McNeill, W., 9

Mechanical machines, 140

Mechanistic world view, 27, 45, 47, 49, 55, 87–88, 92, 259

Mechanization within system, 44, 69–70, 72, 73, 91, 213; and loss of regulability, 70, 213

Meixner, J. R., 142

Mendel's laws, 182

Menninger, Karl, 7, 105, 205, 211

Merloo, Joost, 212

Merton, Robert K., 196

Mesarovic, M. D., 21

Metabolism, 39, 65, 121, 122, 135, 137, 141, 147, 148; self-regulation of, 124, 131; surface law of (Rubner's law), 164–65, 174

Meteorology, 102–103

Metzger, W., 73

Meunier, K., 171

Michelangelo, 192

"Microhistory," 199

Miller, James, 205

Milsum, J. H., 22

Minimum action, principle of, 75, 76, 80

Minsky, Marvin L., 22

Mittasch, A., 71

Mittelstaedt, H., 161

M'Naughten rules, and the criminal, 221

Model and reality, incongruence between, 23–24, 94, 200

Molecular machines, 140

Morchio, R., 102

Morgenstern, O., 15, 22

Morphogenesis, 148–49

Morris, Charles, 90

Moser-Egg, O., 121

Moser, H., 121

Mosteller, F., 49

Motivation research, 116, 189, 191

Müller, I., 181

Mumford, L., 196

Murphy, Gardner, 109

Murray, Henry, 206, 207, 216

Nagel, E., 12

Napoleon, 110–11

Naroll, R. S., 103

Nation, concept of, in U.N., 202

Natural selection, theory of, 47

Neopositivism, 12

Nervous system, new conception of, 106

Net theory approach to systems problems, 21, 90

Netter, H., 102, 158

Neumann, J. von, 15, 22, 25, 27, 246

Newton, Sir Isaac, 186

Nicholas of Cusa, 11, 248

Nietzsche, Friedrich Wilhelm, 187

Nihilism, 187

1984 (Orwell), 10, 52, 118

Nomothetic method in science, 110, 111, 114, 198

"Nothing-but" fallacy in evaluating models, 118–19

Nuttin, Joseph, 216

Oligopoly, law of, and organizations, 48, 104

Onsager, L., 142

Open systems, 11, 13, 21, 23, 32, 39–41, 90, 102–103, 120–124, 141; general characteristics of, 124–32, 141–45; biological applications of concept of, 134–38, 145–49; kinetic theory of, 142, 148, 150; thermodynamic theory of, 142, 144, 148, 150; steady state, false start, and overshoot in, 143 (*fig.*), 160; theory of, as part of general system theory, 149, 154; and cybernetics, 149–50; unsolved problems, 150–53; and steady states, 156–60; in technological chemistry, 122, 142

Operations research, 9, 91, 104

Opinion research, 116

Opler, Marvin, 207

Organic mechanism, philosophy of, 12

Organism: concept of, 67–68; and personality, 105, 208; as open system, 120–24, 134, 153–54; machine model of, and its limitations, 139–41; as active system, 208–10; see also Living organism

Organismic analogy in sociology and history, 116–18

Organismic biology, 6, 12, 31, 88, 89, 102–103, 205, 208, 259

Organismic psychology, 193

Organismic revolution, 186–88

Organismic theory: of personality, 105, 208; of sociology and history, 202–03

Organization: concept of, 46–48, 92, 94, 253; general theory of, 34; characteristics of, 47; aspects of, not subject to quantitative interpretation, 21, 47; Theory of (formal) organizations, 9, 10; Iron Laws of, 47–48, 53, 104; law of optimum size, 48, 104; law of oligopoly, 48, 104; ultimate precept of theory of, 52–53

Organizational Revolution, The (Boulding), 47

Organized complexity, problems of, 34, 93

Ortega y Gasset, José, 118

Orwell, George, 52

Osterhout, W. J. V., 134

Oxenstierna, Count Axel Gustaffson, 116

Paracelsus, 11

Parallelism of cognitive principles in different fields, 31

Pareto's law in sociology, 65, 82

Parseval, August von, 13

Parsons, Talcott, 196

Patten, B. C., 102

Patterns of Culture (Benedict), 201

Pavlov, Ivan Petrovich, 189, 216

Permeability (cell), and open system, 134–35

Personality: theory of, 105–109, 187, 193; and organism, 105, 208; and environmentalism, 189–90; splitting of, 215; as system, 219

Perspectivism, 49, 247

Peterson, R. E., 196

Pharmacodynamics, basic laws of, 56, 138, 148

Phenomenological principles of life, 152

Physical chemistry: trend toward generalized theories in, 32; kinetics and equilibria in chemical systems, 120–22; of enzyme reactions, 143, 147; see also Open systems

Physicalism, 88

Physics: impact of systems thinking on, 5–6; modern developments in 5–6, 30–31; generalized theories in, 32; and theory of unorganized complexity, 34, 93

Physiological clock, 230

Piaget, Jean, 6, 193, 194, 207, 212, 221

Picard, E., 133

Pirozynski, W. J. P., 171

Pitts, W. H., 25

Plato, 52, 235, 240

Politicians, and application of systems approach, 4

Population: periodic cycles in, 48; Malthusian law of, 48, 62, 104; Verhulst law of growth of, 62

Population dynamics, 32, 102, 103–104, 138

Population explosion, 118

Positivism, 94

Pötzl, Otto, 13

Pribram, K. H., 211n.

Prigogine, I., 103, 142, 144, 151, 231

Prigogine's Theorem, 151

Psychiatry: systems concepts in, 208–220; increasing interest in general system theory, 7, 193, 205; modern trends in, 193–94; physico-psycho-sociological framework of, 217

Psychoanalysis, 7, 24, 107, 187, 190

Psychological technology, 52

Psychology: application of G. S. T., 6, 105–107, 220; trends in, 31, 193–94; developmental, 194; quandary of modern psychology, 190, 205–207; holistic orientation in, 193; re-orientation of, 193; see also Psychiatry

Pumpian-Mindlin, Eugene, 205

Purposiveness, 216, 258; see also Behavior

Püter, A., 137, 171

Quantum physics, 31
Quastler, H., 22
Queuing theory approach to systems problems, 23

Rameaux, 164
Rapaport, D., 107, 205
Rapoport, A., 15, 19, 21, 25, 100, 101, 104, 113, 114, 117, 199
Rashevsky, N., 21, 113, 133, 134, 137, 144
Rationality, principle of, 115; and human behavior, 115–16
Reafferenzprinzip (Holst), 16
Reductionism, 48, 49, 86–87, 247
Regression, in psychopathology, 214–15
Reichenbach, Hans, 13
Riegl, A., 232
Reik, H. G., 142, 151
Reiner, J. H., 144
Relativity, theory of, 99, 225, 226, 247
Rensch, B., 153
Repge, R., 26
Rescigno, A., 21, 144, 147
Research, *see* General systems research; Motivation research; Operations research; Opinion research
Responsibility, moral and legal question of, 221
Revolt of the Masses (Ortega y Gasset), 118
Richardson, Lewis F., 104, 114, 195
Rise of the West (McNeill), 9
Rittenberg, D., 175
Robot model of human behavior, 188, 190–91, 194, 205, 206
Rogers, Carl R., 207
Roman Empire, decay of, 203
Rosen, R., 21
Rostovtzeff, Michael Ivanovich, 203
Rothacker, Erich, 213
Roux, Wilhelm, 66
Royce, Joseph R., 215
Rubner's law (surface law of metabolism), 164–65, 174
Ruesch, J., 10
Russell, Bertrand, 67, 68

Sarrus, P. F., 164
Schachtel, E. G., 194
Schaffner, Kenneth F., 12
Schaxel, J., 13, 231
Scher, Jordan, 216, 220

Schiller, Claire, 209, 215
Schlick, Moritz, 12, 77
Schönheimer, R., 160
Schrödinger, Erwin, 98, 144
Schulz, G. V., 148, 151
Science: and evolving of similar problems and conceptions in widely different fields, 30–31; and basic problem of general theory of organization, 34; unity of, 48–49, 86–88, 251–53; and society, 51–52; isomorphism in, 80–86; classical, limitations of, 92–93; generalization of basic concepts in, 94
Sciences of man, system concept in, 186–204
Scientific generalists, production of, 49–51
Scientific revolutions, 17–18, 201
Scott, W. G., 9
Second Industrial Revolution, 4, 16
Segre, G., 21, 144
Segregation within system, 68–69, 70, 71
Self-controlling machines, development of, 3, 15, 140
Self-realization as human goal, 109
Self-restoring tendencies of organismic systems, 27
Selye, H., 192
Senses Considered as Perceptual Systems, The (Gibson), 211n.
Sensory physiology, 148
Servomechanism theory in technology, 22, 78, 140
Set theory approach to systems problems, 21
Shannon, Claude, 15, 22, 100
Shannon's Tenth Theorem, 98
Shaw, Leonard, 17
Simon, H. A., 19, 29
Skinner, B. F., 189, 214
Skrabal, A., 56
Skramlik, E. von, 229
Smith, Vincent E., 12
Social phenomena, statistical regularities and laws in, 199
Social sciences: systems perspective in, 7–8; and socio-cultural systems, 7–9, 196, 197, 200–202; development of new concepts in, 31; broad application of term, 194–195; systems in, 194–197

Society for Empirical Philosophy, Berlin; group of, 12–13

Society for General Systems Research, 15

Society, human, *see* Human society

Socio-cultural systems, and social sciences, 7–9, 194–197, 200–202

Sociological technology, 51–52

Sociology, organismic theory of, 202–203

Sorokin, P. A., 7, 10, 195, 196, 198, 200, 201, 207

Space, conquest of, 187

Specialization in modern science, 30

Spemann, Hans, 72

Spengler, Oswald, 8, 12, 110, 112, 116, 117, 118, 198, 199, 200, 201, 202, 203, 223, 233–34, 237, 249

Spiegelman, S., 56, 65

Sprinson, D. B., 175

Stagner, Ross, 211

Static teleology, 77

Statistical laws, and human society, 116, 198

Steady state in organism, 41, 125, 126, 127, 134, 143, 148, 156–60, 209; definition of, 129–31

Stein-Beling, J. von, 230

Stimulus-response (S-R) scheme, 107, 188–89, 191, 193, 209

Stoward, P. J., 151

Stress, 192

Study of History, A (Toynbee), 200

Summative characteristics in complex, 54–55, 91

Summativity, 67; in mathematical sense, 68

Surface law of metabolism, 164–65, 174

Symbolic activities, 215–18, 252

System: as key concept in scientific research, 9; defined as complex of interacting elements, 19, 38, 55–56, 83–84; mathematical definition, 56–60 (*fig.*); active, 150; definition as machine with input, 97

System-theoretical re-orientation in sciences of man, 192–94

Systems engineering, 3–4, 91, 104–105

Systems science, approaches and aims in, 3–10, 89–94; history of, 10–17; trends in, 17–29; mathematical approaches to, 19–23; *see also* General system theory

Syz, Hans, 205

Szent-Györgyi, A., 5

Tales of Hoffman (Offenbach), 140

Tanner, James, 221

Technology: developments in contemporary, 3–5, 9–10, 12, 187, 204; sociological, 51–52; psychological, 52

Teleology, 45–46, 92; static, 77; dynamic, 78–79; *see also* Purposiveness, Directiveness

Theoretical Biochemistry (Netter), 158

Theoretische Biologie (Bertalanffy), 13

"Theory of Colors" (Goethe), 249

Thermodynamics, 13, 39, 140, 141, 143, 150, 151, 156, 159, 245; irreversible, 13, 94, 131, 142, 148, 151, 152, 153, 159, 163; second principle of, 30, 34, 39, 40, 47, 93, 102, 125, 143–44, 159

Thompson, J. W., 102

Thumb, Norbert, 216

Toch, Hans, 211

Tolstoy, Leo, 111

Topology, 90, 237

Totalitarianism, systems of modern, 52

Toynbee, Arnold, 8, 110, 112, 117, 118, 198, 199, 200, 201, 203, 249

Transport, active, 148

"Tree and the Candle, The" (Koestler), 29

Tribiño, S. E. M. G. de, 12

Trincher, K. S., 152

Tschermak, A., 124

Tukey, F., 49

Turing machine, 22, 25, 141

Turnover rates, 145–147 (*tables*)

Uexküll, Jacob von, 194, 227, 228, 229, 230, 235, 239, 241, 242, 244

Umrath, K., 137

Unconscious drives, 116

Underdeveloped nations, emergence of, 118

Ungerer, E., 12

Unified Theory of Human Behavior, 7

Unity of science, and general system theory, 48–49, 86–88, 259

Unorganized complexity, 34

erbal models in systems theory, 24
erhulst law of population growth, 62
ickers, Sir Geoffrey, 117
ico, Giovanni Battista, 11, 110, 117, 198, 199
ienna Circle, 12
italism, 40, 66, 77, 79, 124, 133, 141, 144, 145
olterra, V., 32, 47, 48, 56, 57, 65, 66, 76, 80, 101, 103, 104, 113, 133, 134, 138

agner, Richard, 16, 101, 161
ahl, O., 230
atson, John B., 189
att, K. E. F., 104
eaver, Warren, 15, 22, 34, 93, 100
eiss, P., 27, 100
erner, G., 56, 148
erner, Heinz, 193, 194, 207, 209, 211, 212
estern civilization, worldwide expansion of, 118
hatmough, J., 248
hite, K., 102
hitehead, A. N., 12, 47, 208

Whittacker, R. H., 41, 102
Wholeness, 31, 55; general science of, 37, 45, 253; and sum, contrast between, and evolution, 70, 94; *see also* Organization
Whorf, B. L., 194, 213, 222, 223, 224, 225, 238; hypothesis of linguistic determination of categories of cognition, 222–27
Whyte, Lancelot, 214, 221
Wiener, Norbert, 15, 44, 78, 101, 161
Winsor, C., 49
Wolfe, Harry B., 4
Woodger, J. H., 29, 208
World, conception of: as chaos, 187; as organization, 188
Worringer, W., 232

Zacharias, J. R., 93
Zeiger, K., 158
Zerbst, E., 144, 147
Zopf, G. W., Jr., 163
Zucker, L., 180
Zucker, T. F., 180
Zwaardemaker, H., 121